Algebra 1

LARSON
BOSWELL
KANOLD
STIFF

Applications • Equations • Graphs

Chapter 10
Resource Book

The Resource Book contains the wide variety
of blackline masters available for Chapter 10.
The blacklines are organized by lesson. Included
are support materials for the teacher as well as
practice, activities, applications, and assessment
resources.

McDougal Littell
A HOUGHTON MIFFLIN COMPANY
Evanston, Illinois • Boston • Dallas

Contributing Authors

The authors wish to thank the following individuals for their contributions to the Chapter 10 Resource Book.

Rita Browning
Linda E. Byrom
José Castro
Christine A. Hoover
Carolyn Huzinec
Karen Ostaffe
Jessica Pflueger
Barbara L. Power
James G. Rutkowski
Michelle Strager

Pages 48, 62, 117: Excerpted and adapted from the World Book Encyclopedia. Copyright © 2000 World Book, Inc. By permission of the publisher. www.worldbook.com

ISBN: 0-618-02048-9

3456789-CKI- 04 03 02 01

Contents

10 *Polynomials and Factoring*

Contents

Contents

Descriptions of Resources

This Chapter Resource Book is organized by lessons within the chapter in order to make your planning easier. The following materials are provided:

Tips for New Teachers These teaching notes provide both new and experienced teachers with useful teaching tips for each lesson, including tips about common errors and inclusion.

Parent Guide for Student Success This guide helps parents contribute to student success by providing an overview of the chapter along with questions and activities for parents and students to work on together.

Prerequisite Skills Review Worked-out examples are provided to review the prerequisite skills highlighted on the Study Guide page at the beginning of the chapter. Additional practice is included with each worked-out example.

Strategies for Reading Mathematics The first page teaches reading strategies to be applied to the current chapter and to later chapters. The second page is a visual glossary of key vocabulary.

Lesson Plans and Lesson Plans for Block Scheduling This planning template helps teachers select the materials they will use to teach each lesson from among the variety of materials available for the lesson. The block-scheduling version provides additional information about pacing.

Warm-Up Exercises and Daily Homework Quiz The warm-ups cover prerequisite skills that help prepare students for a given lesson. The quiz assesses students on the content of the previous lesson. (Transparencies also available)

Activity Support Masters These blackline masters make it easier for students to record their work on selected activities in the Student Edition.

Alternative Lesson Openers An engaging alternative for starting each lesson is provided from among these four types: *Application, Activity, Graphing Calculator,* or *Visual Approach.* (Color transparencies also available)

Graphing Calculator Activities with Keystrokes Keystrokes for four models of calculators are provided for each Technology Activity in the Student Edition, along with alternative Graphing Calculator Activities to begin selected lessons.

Practice A, B, and C These exercises offer additional practice for the material in each lesson, including application problems. There are three levels of practice for each lesson: A (basic), B (average), and C (advanced).

Contents

Reteaching with Practice These two pages provide additional instruction, worked-out examples, and practice exercises covering the key concepts and vocabulary in each lesson.

Quick Catch-Up for Absent Students This handy form makes it easy for teachers to let students who have been absent know what to do for homework and which activities or examples were covered in class.

Cooperative Learning Activities These enrichment activities apply the math taught in the lesson in an interesting way that lends itself to group work.

Interdisciplinary Applications/Real-Life Applications Students apply the mathematics covered in each lesson to solve an interesting interdisciplinary or real-life problem.

Math and History Applications This worksheet expands upon the Math and History feature in the Student Edition.

Challenge: Skills and Applications Teachers can use these exercises to enrich or extend each lesson.

Quizzes The quizzes can be used to assess student progress on two or three lessons.

Chapter Review Games and Activities This worksheet offers fun practice at the end of the chapter and provides an alternative way to review the chapter content in preparation for the Chapter Test.

Chapter Tests A, B, and C These are tests that cover the most important skills taught in the chapter. There are three levels of test: A (basic), B (average), and C (advanced).

SAT/ACT Chapter Test This test also covers the most important skills taught in the chapter, but questions are in multiple-choice and quantitative-comparison format. (See *Alternative Assessment* for multi-step problems.)

Alternative Assessment with Rubrics and Math Journal A journal exercise has students write about the mathematics in the chapter. A multi-step problem has students apply a variety of skills from the chapter and explain their reasoning. Solutions and a 4-point rubric are included.

Project with Rubric The project allows students to delve more deeply into a problem that applies the mathematics of the chapter. Teacher's notes and a 4-point rubric are included.

Cumulative Review These practice pages help students maintain skills from the current chapter and preceding chapters.

Tips for New Teachers

For use with Chapter 10

LESSON 10.1

INCLUSION Students with learning disabilities might need a concrete model to understand abstract concepts. These students will benefit from using algebra tiles to model and to manipulate polynomials. The activity on page 575 shows how to add polynomials using algebra tiles. Make sure students understand *zero pairs* and the ideas behind combining "like tiles" before attempting to use this model for addition. The algebra tiles can also be used to subtract polynomials by "removing" tiles from the board. If it becomes necessary, zero pairs can be added to the original polynomial before the subtraction takes place.

COMMON ERROR If some of your students still make errors combining like terms, use the algebra tiles to show them that x and x^2 are not the same thing. Students are less likely to combine terms with different powers of a variable when they identify each power with a different visual model, a different tile.

LESSON 10.2

TEACHING TIP For those students who get lost multiplying polynomials using the distributive property, you can try this other method. Draw a big rectangle divided into smaller rectangles. Each polynomial represents the length of a side of the big rectangle. Each term of the polynomials represents the length of a side of one of the small rectangles. The product of the polynomials is the area of the big rectangle, which can be found by adding up the areas of the smaller rectangles. For instance, to multiply $(4x - 3) \cdot (2x + 5)$ show:

	$4x$	-3
$2x$	$8x^2$	$-6x$
$+5$	$20x$	-15

Add up the areas and group the like terms to get the answer, $8x^2 + 14x - 15$.

TEACHING TIP Some students learn how to use FOIL to multiply binomials but do not know what to do when one of the polynomials they need to multiply is a trinomial. These students do not understand the distributive property. Show students how to use this property when the polynomials have more than two terms.

LESSON 10.3

COMMON ERROR Even after completing this lesson there will be students who will equate $(x + 4)^2$ and $x^2 + 16$. These same students would probably get the correct answer if they rewrote the problem as a product of two identical binomials and then used FOIL or any other method to multiply them out. Although the goal of this lesson is to identify and use the special product patterns, those students who do not use the patterns correctly might be better off finding or at least checking their answers by using FOIL.

LESSON 10.4

COMMON ERROR Some students might erroneously think that $x = 2$ is one of the solutions of the equation $(3x - 2)(x + 1) = 0$. Even if the linear factors of the polynomial are simple, emphasize the need to use the zero-product property to set each factor equal to zero. Then solve each of the resulting equations to find the solutions to the original equation.

LESSON 10.5

INCLUSION Students with learning disabilities will benefit from using the algebra tiles to model factoring trinomials. Make sure to complete the activities on pages 603 and 610 before Lessons 10.5 and 10.6 respectively. While the tiles are a good model to factor trinomials when the coefficients of all terms are positive, they do not work as well when some of the terms have negative coefficients. Consider using the tiles as a means to build students' understanding of the factoring process, but eventually students will have to move away from tiles and use paper-and-pencil methods.

TEACHING TIP Let your students work with the algebra tiles until someone discovers a rule to factor $x^2 + bx + c$ without using the tiles. With enough time and examples the students are bound to find the pattern.

TEACHING TIP In this lesson, students will discover that many polynomials cannot be factored using integers. What does that mean in terms of solutions and x-intercepts? Show students that some of these non-factorable polynomials, such as $x^2 + x + 2$, have no real solutions. The graphs of

LESSON 10.5 (CONT.)

these polynomials do not cross the *x*-axis. Other polynomials, such as $x^2 + 3x - 6$, cannot be factored because their solutions are not integers. Nevertheless, the polynomials *do* have real solutions and, therefore, their graphs cross the *x*-axis. A graphing calculator might be helpful to demonstrate the above information to your students.

LESSON 10.6

TEACHING TIP The procedure outlined in this lesson is probably the fastest method to factor trinomials of the form $ax^2 + bx + c$. However, some students may not like this method because it may require much trial and error and record keeping. A less creative, more algorithmic method is the *British method*, outlined on page 631. If you decide to teach the *British method* to your students, you must first show them how to use the *grouping* technique shown in Lesson 10.8.

COMMON ERROR Students might claim that a problem is not factorable without trying all the possible pairs of factors. Remind students that they can use the discriminant to decide whether the polynomial is factorable. You can ask your students to calculate and show the value of the discriminant whenever they claim that a polynomial cannot be factored.

LESSON 10.7

TEACHING TIP Remind students that they can always go back to the techniques of Lessons 10.5

and 10.6 if they find it easier than finding patterns. If they choose to do so, students might need to rewrite a difference of two squares pattern, such as $x^2 - 9$, as $x^2 + 0 \cdot x - 9$ to be able to factor it.

TEACHING TIP While factoring a quadratic equation is a very powerful method to find its solutions, it might not always be the fastest one. In Example 5 on page 621, students could solve the equation $150 \cdot C^2 = 600$ by dividing both sides by 150 and then taking the square root of both sides. Students should be able to choose the most efficient method to solve each problem.

LESSON 10.8

COMMON ERROR Students who did not do well with powers will need special attention when learning how to find the GCF and how to decide what is left inside the parentheses after they factor the GCF out. You might have to review powers and properties of exponents.

COMMON ERROR Some students might incorrectly use the zero-product property when the expression is factored but not equal to zero. For instance, in Example 8 on page 627, the algebraic model is $x(x - 1)(x + 4) = 12$. Since the left side of the equation is already factored, students might think that the solutions are 0, 1, and –4. Remind your students that the zero-product property can only be used when the polynomial is factored *and* its product is equal to zero.

Outside Resources

BOOKS/PERIODICALS

Binder, Margery. "A Calculator Investigation of an Interesting Polynomial." *Mathematics Teacher* (October 1995); pp. 558–560.

ACTIVITIES/MANIPULATIVES

Algebra Lab Gear. Manipulatives for teaching algebra concepts. Creative Publications, 1990.

SOFTWARE

Harvey, Wayne, Judah Schwartz, and Michael Yerushalmy. *Visualizing Algebra: The Function Analyzer.* Scotts Valley, CA; Sunburst.

VIDEOS

Apostel, Tom. *Polynomials.* Reston, VA; NCTM.

CHAPTER

10

Chapter Support

NAME _____ DATE _____

Parent Guide for Student Success

For use with Chapter 10

Chapter Overview One way that you can help your student succeed in Chapter 10 is by discussing the lesson goals in the chart below. When a lesson is completed, ask your student to interpret the lesson goals for you and to explain how the mathematics of the lesson relates to one of the key applications listed in the chart.

Lesson Title	Lesson Goals	Key Applications
10.1: Adding and Subtracting Polynomials	Add and subtract polynomials. Use polynomials to model real-life situations.	• Mounting a Photo • Building a House • Energy Use
10.2: Multiplying Polynomials	Multiply two polynomials. Use polynomial multiplication in real-life situations.	• Carpentry • Football Field • Videocassette Sales
10.3: Special Products of Polynomials	Use special product patterns for the product of a sum and a difference and for the square of a binomial. Use special products as real-life models.	• Punnett Squares • Genetics • Investment Value
10.4: Solving Polynomial Equations in Factored Form	Solve a polynomial equation in factored form. Relate factors and x-intercepts.	• Arecibo Observatory • Gateway Arch • Barringer Meteor Crater
10.5: Factoring $x^2 + bx + c$	Factor a quadratic expression of the form $x^2 + bx + c$. Solve quadratic equations by factoring.	• Landscape Design • Making a Sign • The Taj Mahal
10.6: Factoring $ax^2 + bx + c$	Factor a quadratic expression of the form $ax^2 + bx + c$. Solve quadratic equations by factoring.	• Cliff Diving • Gymnastics • Warp and Weft
10.7: Factoring Special Products	Use special product patterns to factor quadratic polynomials. Solve quadratic equations by factoring.	• Block and Tackle • Safe Working Load • Pole Vaulting
10.8: Factoring Using the Distributive Property	Use the distributive property to factor a polynomial. Solve polynomial equations by factoring.	• Vertical Motion and Gravity • Packaging

Test-Taking Strategy

Read test questions carefully to be sure that you **answer the right question.** Watch out for questions that involve more than one step. Ask your student to find an example of a problem in the chapter that involves two or more steps. Have your student give the answer to a preliminary step and then give the correct final answer.

NAME _____ DATE _____

Parent Guide for Student Success

For use with Chapter 10

Key Ideas Your student can demonstrate understanding of key concepts by working through the following exercises with you.

Lesson	Exercise
10.1	Find the sum and the difference of $(4x^3 - 2x + 7)$ and $(2x^3 + 5x - 9)$.
10.2	The number of calendars a class can sell is given by $50 + 4x$. Each calendar is sold for $0.15x + 3$ dollars. Find a model for the class's total revenue from selling calendars as a quadratic trinomial.
10.3	Show how to use mental math and the Sum and Difference Pattern to find $38 \cdot 42$.
10.4	A highway arch can be modeled in feet by $y = -0.125(x - 12)(x + 12)$. How wide is the arch at the base? How high is the arch?
10.5	Solve $x^2 - x - 42 = 0$.
10.6	A competitive diver jumps from a 10-foot high platform with an initial velocity of 6 feet per second. The diver's height h, t seconds after she jumps, can be modeled by $h = -16t^2 + 6t + 10$. How long does it take the diver to hit the water?
10.7	The area of a square garden is modeled by $A = x^2 - 24x + 144$ square feet. Use the variable x to express the dimensions of the garden.
10.8	Factor the expression completely. $4x^2y - 44xy + 72y$

Home Involvement Activity

You will need: A tape measure, a rectangular room

Directions: Measure the length and width of a rectangular room in your home, to the nearest foot. Suppose you want to buy an oriental carpet to fit in the room with a space x feet wide on all four sides. Find a model for the area of the rug. Write it as a quadratic trinomial. Use the model to find the cost of the carpet if $x = 3$ and carpet costs $7.50 a square foot.

Answers

10.1: $6x^3 + 3x - 2$; $2x^3 - 7x + 16$ **10.2:** $R = 0.6x^2 + 19.5x + 150$
10.3: $38 \cdot 42 = (40 - 2)(40 + 2) = 1600 - 4 = 1596$ **10.4:** 24 ft; 18 ft **10.5:** 7, -6 **10.6:** 1 sec
10.7: $(x - 12)$ ft by $(x - 12)$ ft **10.8** $4y(x - 9)(x - 2)$

Algebra 1
Chapter 10 Resource Book

Prerequisite Skills Review

For use before Chapter 10

EXAMPLE 1 *Simplifying by Combining Like Terms*

Combine like terms.

$6x - 4(7 + x)$

SOLUTION

$$
\begin{aligned}
6x - 4(7 + x) &= 6x + (-4)(7 + x) && \text{Rewrite as an addition expression.}\\
&= 6x + [(-4)(7) + (-4)(x)] && \text{Distribute the } -4.\\
&= 6x + (-28) + (-4x) && \text{Multiply.}\\
&= 2x - 28 && \text{Combine like terms and simplify.}
\end{aligned}
$$

Exercises for Example 1

Combine like terms.

1. $(5 + 10x)2$ **2.** $-\frac{1}{4}(t - 16)$ **3.** $-(y - 6) - y$

4. $7s + 2(s - 10)$ **5.** $3.2(7.2 + x) + x$ **6.** $3x^2 - x(3 + 2x)$

EXAMPLE 2 *Multiplying Exponential Expressions*

Simplify the expression.

a. $(-2st^3)^3(-s^4)^5$ **b.** $(5x^3y^4)^2 \cdot x^6$

SOLUTION

a.
$$
\begin{aligned}
(-2st^3)^3(-s^4)^5 &= (-2)^3(s)^3(t^3)^3(-s^4)^5 && \text{Raise each factor to a power.}\\
&= -8 \cdot s^3 \cdot t^9 \cdot (-s^{20}) && \text{Evaluate each power.}\\
&= 8s^{23}t^9 && \text{Simplify.}
\end{aligned}
$$

b.
$$
\begin{aligned}
(5x^3y^4)^2 \cdot x^6 &= 5^2(x^3)^2(y^4)^2 \cdot x^6 && \text{Raise each factor to a power.}\\
&= 25 \cdot x^6 \cdot y^8 \cdot x^6 && \text{Evaluate each power.}\\
&= 25x^{12}y^8 && \text{Simplify.}
\end{aligned}
$$

Exercises for Example 2

Simplify the expression.

7. $x^2 \cdot y^5 \cdot x^4 \cdot y^{12}$ **8.** $(-4ab^4)(3a^6b^9)$

9. $(4s^5t^9)^3 \cdot 3t^4$ **10.** $\left(\frac{1}{2}x^4y^6\right)^5\left(x^4y\right)^2$

NAME _____ DATE _____

Prerequisite Skills Review

For use before Chapter 10

EXAMPLE 3 *Finding the Number of Solutions*

Tell if the equation has two solutions, one solution, or no solution.

a. $2x^2 - 4x + 5 = 0$

b. $-5x^2 - 7x + 10 = 0$

SOLUTION

a. $b^2 - 4ac = (-4)^2 - 4(2)(5)$ Substitute 2 for a, -4 for b, 5 for c.

$\qquad\qquad = 16 - 40$ Simplify.

$\qquad\qquad = -24$ Discriminant is negative.

The discriminant is negative, so the equation has no solution.

b. $b^2 - 4ac = (-7)^2 - 4(-5)(10)$ Substitute -5 for a, -7 for b, 10 for c.

$\qquad\qquad = 49 + 200$ Simplify.

$\qquad\qquad = 249$ Discriminant is positive.

The discriminant is positive, so the equation has two solutions.

Exercises for Example 3

Tell if the equation has two solutions, one solution, or no solution.

11. $2x^2 - 4x + 10 = 0$

12. $x^2 - 2x + 1 = 0$

13. $-3x^2 - 8x + 1 = 0$

14. $x^2 - 4x + 16 = 0$

Strategies for Reading Mathematics

For use with Chapter 10

Strategy: Reading Specialized Vocabulary

Every occupation, sport, and hobby has its own specialized vocabulary that lets enthusiasts talk to each other using precise terms that each person understands. Mathematics has its own specialized vocabulary. Making yourself familiar with the language will make reading algebra much easier.

Through repeated use, some mathematical terms become familiar enough for everyday use.

An expression which is the sum of terms of the form ax^k where k is a nonnegative integer is a **polynomial.** Polynomials are usually written in **standard form,** which means that the terms are placed in descending order, from largest degree to smallest degree.

Leading Coefficient ↘ ↙ Degree   ↙ Constant term

Polynomial in standard form: $2x^3 + 5x^2 - 4x + 7$

The **degree** of each term of a polynomial is the exponent of the variable. The **degree of a polynomial** is the largest degree of its terms. When a polynomial is written in standard form, the coefficient of the first term is the **leading coefficient.**

A word may have one or more meanings in everyday speech and others in mathematics.

Specialized language may be defined in words or by example.

STUDY TIP
Look It Up

You may come across a term whose meaning you have forgotten. If you cannot determine the meaning from context clues in the text or from the examples, don't just skip the word. Find the meaning in the Glossary or in a dictionary. Then reread the sentence.

STUDY TIP
Speak and Write Like a Mathematician

The more you use mathematical terms as you speak and write, the more familiar they will become. Try to use precise terms as you explain mathematical processes, so that the terms become as familiar as everyday speech.

Questions

Use the example above for Questions 1–3.

1. What are some of the meanings of the word *degree* other than the one given here? When you come upon the word *degree* as you are reading, how can you tell which meaning is intended?

2. Why is understanding the term *nonnegative integer* an important part of knowing what a polynomial is? Give an example of an expression that is *not* a polynomial, based on this restriction.

3. Explain why the terms *integer*, *variable*, and *coefficient* in the text above are neither defined nor explained by example.

NAME _____ DATE _____

Strategies for Reading Mathematics

For use with Chapter 10

Visual Glossary

The Study Guide on page 574 lists the key vocabulary for Chapter 10 as well as review vocabulary from previous chapters. Use the page references on page 574 or the Glossary in the textbook to review key terms from prior chapters. Use the visual glossary below to help you understand some of the key vocabulary in Chapter 10. You may want to copy these diagrams into your notebook and refer to them as you complete the chapter.

GLOSSARY

polynomial (p. 576) An expression which is the sum of terms of the form ax^k where k is a nonnegative integer.

factor a quadratic expression (p. 604) To write a quadratic expression as the product of two linear expressions.

FOIL pattern (p. 585) A pattern used to multiply two binomials. Multiply the First, Outer, Inner, and Last terms.

factor completely (p. 625) To write a polynomial as the product of:
• monomial factors
• prime factors with at least two terms

factored form (p. 597) A polynomial that is written as the product of two or more linear factors.

zero-product property (p. 597) A property that states that the product of two factors is zero, only when at least one of the factors is zero.

Representing the Factors of Polynomials

Once you have determined the factors of a polynomial, you can check your work algebraically using the FOIL pattern, or you can model with or sketch algebra tiles to see that your work is correct.

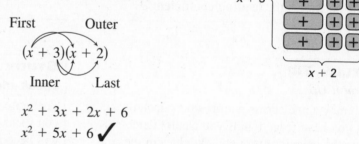

$$x^2 + 5x + 6 = (x + 2)(x + 3)$$

factor a quadratic expression

Check: Use the FOIL pattern.

First   Outer

$$(x + 3)(x + 2)$$

Inner   Last

$$x^2 + 3x + 2x + 6$$
$$x^2 + 5x + 6 \checkmark$$

Representing the Zero-Product Property

The zero-product property is useful in solving quadratic equations and sketching graphs of quadratic functions.

Solve $6x^2 - 8x - 8 = 0$.

$$6x^2 - 8x - 8 = 0$$
$$2(3x^2 - 4x - 4) = 0$$
$$2(3x + 2)(x - 2) = 0 \quad \longleftarrow \text{ Factor completely.}$$
$$(3x + 2)(x - 2) = 0$$

$(3x + 2)(x - 2) = 0$ so $3x + 2 = 0$ or $x - 2 = 0$. $\longleftarrow$ zero-product property

$$3x + 2 = 0 \qquad\qquad x - 2 = 0$$
$$x = -\tfrac{2}{3} \qquad\qquad x = 2$$

The solutions are $-\tfrac{2}{3}$ and 2.

Lesson Plan

1-day lesson (See *Pacing the Chapter*, TE pages 572C–572D)   **For use with pages 575–583**

GOALS   1. **Add and subtract polymonials.**
2. **Use polynomials to model real-life situations.**

State/Local Objectives _____

_____

✓ Check the items you wish to use for this lesson.

STARTING OPTIONS
_____ Prerequisite Skills Review: CRB pages 5–6
_____ Strategies for Reading Mathematics: CRB pages 7–8
_____ Warm-Up or Daily Homework Quiz: TE pages 576 and 560, CRB page 11, or Transparencies

TEACHING OPTIONS
_____ Motivating the Lesson: TE page 577
_____ Concept Activity: SE page 575; CRB page 12 (Activity Support Master)
_____ Lesson Opener (Application): CRB page 13 or Transparencies
_____ Graphing Calculator Activity with Keystrokes: CRB page 14
_____ Examples 1–6: SE pages 576–579
_____ Extra Examples: TE pages 577–579 or Transparencies; Internet
_____ Technology Activity: SE page 583
_____ Closure Question: TE page 579
_____ Guided Practice Exercises: SE page 579

APPLY/HOMEWORK
Homework Assignment
_____ Basic   20–62 even, 65, 66, 70, 75, 80, 85
_____ Average   20–62 even, 63–66, 70, 75, 80, 85
_____ Advanced   20–62 even, 63–66, 70–72, 75, 80, 85

Reteaching the Lesson
_____ Practice Masters: CRB pages 15–17 (Level A, Level B, Level C)
_____ Reteaching with Practice: CRB pages 18–19 or Practice Workbook with Examples
_____ Personal Student Tutor

Extending the Lesson
_____ Applications (Interdisciplinary): CRB page 21
_____ Challenge: SE page 582; CRB page 22 or Internet

ASSESSMENT OPTIONS
_____ Checkpoint Exercises: TE pages 577–579 or Transparencies
_____ Daily Homework Quiz (10.1): TE page 582, CRB page 25, or Transparencies
_____ Standardized Test Practice: SE page 582; TE page 582; STP Workbook; Transparencies

Notes _____

_____

_____

Lesson Plan for Block Scheduling

Half-day lesson (See *Pacing the Chapter*, TE pages 572C–572D)   **For use with pages 575–583**

GOALS   1. **Add and subtract polymonials.**
            2. **Use polynomials to model real-life situations.**

State/Local Objectives _____

_____

_____

✓ **Check the items you wish to use for this lesson.**

STARTING OPTIONS

____ Prerequisite Skills Review: CRB pages 5–6
____ Strategies for Reading Mathematics: CRB pages 7–8
____ Warm-Up or Daily Homework Quiz: TE pages 576 and
      560, CRB page 11, or Transparencies

TEACHING OPTIONS

____ Motivating the Lesson: TE page 577
____ Concept Activity: SE page 575; CRB page 12 (Activity Support Master)
____ Lesson Opener (Application): CRB page 13 or Transparencies
____ Graphing Calculator Activity with Keystrokes: CRB page 14
____ Examples 1–6: SE pages 576–579
____ Extra Examples: TE pages 577–579 or Transparencies; Internet
____ Technology Activity: SE page 583
____ Closure Question: TE page 579
____ Guided Practice Exercises: SE page 579

APPLY/HOMEWORK

Homework Assignment

____ Block Schedule: 20–62 even, 63–66, 70, 75, 80, 85

Reteaching the Lesson

____ Practice Masters: CRB pages 15–17 (Level A, Level B, Level C)
____ Reteaching with Practice: CRB pages 18–19 or Practice Workbook with Examples
____ Personal Student Tutor

Extending the Lesson

____ Applications (Interdisciplinary): CRB page 21
____ Challenge: SE page 582; CRB page 22 or Internet

ASSESSMENT OPTIONS

____ Checkpoint Exercises: TE pages 577–579 or Transparencies
____ Daily Homework Quiz (10.1): TE page 582, CRB page 25, or Transparencies
____ Standardized Test Practice: SE page 582; TE page 582; STP Workbook; Transparencies

Notes _____

_____

_____

CHAPTER PACING GUIDE	
Day	**Lesson**
1	Assess Ch. 9; **10.1 (all)**
2	10.2 (all); 10.3 (begin)
3	10.3 (end); 10.4 (all)
4	10.5 (all)
5	10.6 (all)
6	10.7 (all)
7	10.8 (all)
8	Review/Assess Ch. 10

NAME ⎯⎯⎯⎯⎯⎯⎯⎯⎯⎯⎯⎯⎯⎯⎯⎯⎯ DATE ⎯⎯⎯⎯⎯

WARM-UP EXERCISES

For use before Lesson 10.1, pages 575–583

Give the slope and the *y*-intercept of the graph of the equation.

1. $y = 4x - 2$

2. $y = \dfrac{2}{3}x + \dfrac{4}{3}$

3. $2x - y = 8$

Simplify.

4. $2x + 5 - 7x - 7$

5. $7 - (2x + 8) - 3$

DAILY HOMEWORK QUIZ

For use after Lesson 9.8, pages 554–560

Make a scatter plot of the data. Name the type of model that best fits the data.

1. $\left(1, -\dfrac{15}{4}\right), (-2, -2), \left(2, -\dfrac{31}{8}\right), (-1, -3), (-3, 0), (-4, 4)$

2. $(0, -1.5), (2, 2.5), (-3, 0), (1, 0), (-1, -2), (-2, -1.5)$

3. The ordered pairs give the year and the population in thousands of a town. Write a model that accurately reflects the data.
$(1995, 24.5), (1996, 27.0), (1997, 29.6), (1998, 32.6), (1999, 35.9)$

Algebra 1
Chapter 10 Resource Book

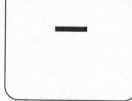
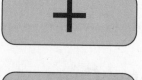
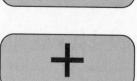

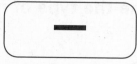

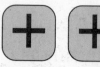

Application Lesson Opener

For use with pages 576–582

In Questions 1 and 2, use the following information.

Ramon has a 3 in. by 5 in. photo. He wants to enlarge both dimensions by the same amount. If x represents the amount of increase, the new dimensions can be written as $(3 + x)$ in. and $(5 + x)$ in.

1. Which expression represents the perimeter of the enlarged photo?

    A. $(3 + x) + (5 + x)$

    B. $(3 + x)(5 + x)$

    C. $(3 + x) + (3 + x) + (5 + x) + (5 + x)$

    D. $2(3 + x)(5 + x)$

2. Explain how to simplify the expression you chose in Question 1. Then simplify.

In Questions 3 and 4, use the following information.

Ramon is going to mat and frame the enlarged photo. The photo will be centered on the mat. The area of the enlarged photo is $15 + 8x + x^2$. The area of the mat for the photo is $40 + 13x + x^2$.

3. Which expression represents the difference of the areas of the mat and the photo?

    A. $(40 + 13x + x^2) - (15 + 8x + x^2)$

    B. $(15 + 8x + x^2) - (40 + 13x + x^2)$

    C. $15 + 8x + x^2 - 40 + 13x + x^2$

    D. $40 + 13x + x^2 - 15 + 8x + x^2$

4. Explain how to simplify the expression you chose in Question 3. Then simplify.

Graphing Calculator Activity Keystrokes

For use with Technology Activity 10.1 on page 583

TI-82

TI-83

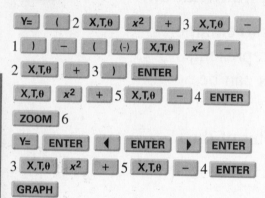

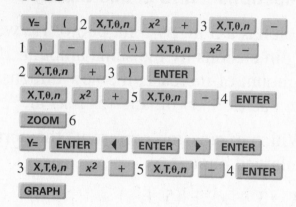

SHARP EL-9600c

CASIO CFX-9850GA PLUS

From the main menu, choose GRAPH.

[( ] [2] [X,θ,T] [x^2] [+] [3] [X,θ,T] [−] [1] [ )] [−]

[( ] [(-)] [X,θ,T] [x^2] [−] [2] [X,θ,T] [+] [3] [ )]

[EXE]

[X,θ,T] [x^2] [+] [5] [X,θ,T] [−] [4] [EXE]

[SHIFT] [F3] [F3]

[EXIT] [F6]

[EXIT] [▲] [F1] [▼]

[3] [X,θ,T] [x^2] [+] [5] [X,θ,T] [−] [4] [EXE]

[F6]

Practice A

For use with pages 576–582

Write the polynomial in standard form.

1. $3x + 4x^2 - 5$

2. $5x^2 + 4 - 3x$

3. $x - 7x^3 + 2$

4. $8 + 2x + 4x^2$

5. $5x^2 + 4x^3 - 2x$

6. $-4x + 7x^4 - 5x^3 + 1$

7. $3x - 7 + 2x^2$

8. $7x - 2$

9. $-x + 2x^2 + x^3 - 2$

Identify the leading coefficient, and classify the polynomial by degree and by number of terms.

10. 14

11. $2x + 3$

12. $-3x^2 + 6x - 2$

13. $x^3 - 5$

14. $1 - x^4$

15. $x^2 + 4x - x^4 + 3x^3 - 8$

16. $2x^2 - 5x + 1$

17. $2 + x^2$

18. $x - x^3 + 3x^2 + 9$

Use a vertical format to add or subtract.

19. $(x^2 + 2x + 7) + (4x^2 + x - 3)$

20. $(5x^2 - 2x + 4) + (-2x^2 + 3x - 1)$

21. $(5n^2 + 2n + 3) - (n + 2)$

22. $(6n^2 + 4n + 6) - (5n^2 + n + 2)$

23. $(2a^3 - 4a^2 + 7) + (-2a^2 + a - 3)$

24. $(4n^2 - 6n + 5) - (8n^2 + n + 3)$

Use a horizontal format to add or subtract.

25. $(x^2 + 2x + 1) + (x - 3)$

26. $(3m^2 + 2m + 1) - (-2m^2 + 4m)$

27. $(7x + 1) - (-x^2 + 3x - 5)$

28. $(5x^2 - 9) + (-3x^2 + 5x + 9)$

29. $(7x^3 - 8x^2 + 4) + (9x^2 + 5x + 2)$

30. $(n^2 - 2n) - (-5n^2 + 3n - 1)$

Use a vertical format or a horizontal format to add or subtract.

31. $(2x^3 + 5x) + (7x^2 - x + 2)$

32. $(-2x^2 - 7x + 3) - (-5x^2 + 3x - 7)$

33. $(3 + 7x) + (13x - 4)$

34. $(x^3 + x^2 + x + 1) + (-x^2 - x - 1)$

35. $(2x^2 + 3) - (6x + 4) + (3x^2 - x)$

36. $(x^2 - x + 3) - (-3x^2 + 5x - 2)$

37. *Photograph Mat* A mat in a frame has an opening for a photograph (see figure). Find an expression for the area of the mat. (Area of opening: $A = \pi ab$).

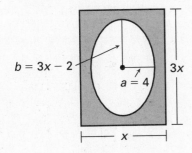

38. *Floor Plan* The first floor of a home has the floor plan shown below. Find an expression in standard form for the area of the first floor.

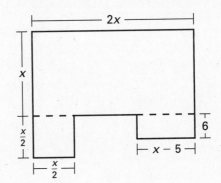

Lesson 10.1

Practice B
For use with pages 576–582

Identify the leading coefficient, and classify the polynomial by degree and by number of terms.

1. 12

2. $5x - 2$

3. $-4x^2 + 7x - 1$

4. $5x^3 - 2$

5. $4 - 3x^4$

6. $x^2 + 5x + 2x^4 + x^3 - 2$

7. $6 - 3x^3$

8. $5x^2 + 3x - 4$

9. $7x - x^5 + 3x^2 + 1$

Use a vertical format to add or subtract.

10. $(3x^2 - 4x + 1) + (-x^2 + x - 9)$

11. $(-8x^2 - 3x + 7) + (-x^3 + 6x^2 - 5)$

12. $(-4x^3 - 8x^2 + 5x + 9) - (6x^2 + 2x - 4)$

13. $(6x - 5) + (4x^3 - 3x + 4)$

14. $(-3 + 4n^2) - (5 - 2n^3)$

15. $(x - 4x^2 + 7) - (-5x^2 + 5x - 3)$

Use a horizontal format to add or subtract.

16. $(x^2 + 1) + (-4x^2 + 5)$

17. $(3x^2 + 4) - (x^2 - 5x + 2)$

18. $(3t^2 - 8t + 2) - (-3t^2 + 5t - 7)$

19. $(9x^3 - 5x^2 + x) + (6x^2 + 5x - 10)$

20. $(5x^2 + x^3 + 6) + (x^2 + 5 - 6x)$

21. $12 - (-5x^2 + x - 7)$

Use a vertical format or a horizontal format to add or subtract.

22. $(5x^3 - 3x) - (7x^2 - 3x + 1)$

23. $(2x^3 + 4) - (-x^2 + 3x)$

24. $(m + 3m^3 - 4m^5) + (2m^3 + 5m^5 - 4)$

25. $(x^2 + 1) + (-3x^2 - 7) - (x^2 + 5)$

26. $(5x^2 + 4) - (3x + 7) + (2x^2 - 1)$

27. $2(x^2 - 4x + 5) - (x^2 + 6x - 1)$

28. *Profit* For 1990 through 2000, the revenue R and cost C of producing a product can be modeled by

$$R = \tfrac{1}{3}t^2 + \tfrac{20}{3}t + 300$$
$$C = \tfrac{1}{15}t^2 + \tfrac{13}{3}t + 200$$

where t is the number of years since 1990. Find a model for the profit P earned from 1990 to 2000. (*Hint:* Profit = Revenue − Cost)

29. *Library Books* For 1990 through 2000, the number of fiction books F (in 10,000s) and nonfiction books N (in 10,000s) borrowed from a library can be modeled by

$$F = 0.01t^2 + 0.09t + 6$$
$$N = 0.004t^2 + 0.06t + 4$$

where t is the number of years since 1990. Find a model for the total number of books borrowed B from the library in a year from 1990 to 2000.

NAME _____ DATE _____

Practice C

For use with pages 576–582

Identify the leading coefficient, and classify the polynomial by degree and by number of terms.

1. -4

2. $3x - 1$

3. $-x^2 + 6x - 2$

4. $7x^3 - 5x$

5. $6 - 2x^3$

6. $x^3 - 5x + 2x^5 - x^4 - 7$

7. $6 + 3x^4$

8. $-x^2 + 6x - 2$

9. $3x - x^4 + 3x^3 - 5x^2$

Use a vertical format to add or subtract.

10. $(2a^2 - 4a + 3) + (6a^2 + 4a - 3)$

11. $(-4x^3 - 7x + 5) + (-x^2 + 6x - 1)$

12. $(-2x^3 + 3x^2 + x + 2) - (x^2 - x + 4)$

13. $(9x - 2) + (2x^4 - 5x + 1)$

14. $(7m^2 - 3m + 8) - (-3m^2 - 6m + 1)$

15. $(-3 + 2n^2 + 5n^5) - (4 - n^3 + 2n^2 + n^5)$

Use a horizontal format to add or subtract.

16. $(5x^2 + 2x - 1) + 8x^2$

17. $(5n^2 + 9) - (n^2 - 8n - 5)$

18. $(n^2 - 6n) + (-2n^2 + 5n + 2)$

19. $(5t^3 - 2t^2 + t) - (-4t^3 + t^2 + 3)$

20. $(7x^2 + x^3 + 9) + (4x^2 + 2 - 5x)$

21. $(6x - 3x^2 + 1) - (9x - 4 - 3x^2)$

Use a vertical format or a horizontal format to add or subtract.

22. $(3n^3 - 5n) - (2n^2 - 4n + 7)$

23. $(x^3 + 4x) - (2x - x^2)$

24. $(3m + m^3 - 2m^5) + (7m^3 + m^5 - 1)$

25. $(x^2 + 1) + (x^2 - 1) - (x^2 + 1)$

26. $(6x - 5) - (8x + 15) + (3x - 4)$

27. $(2x^2 + 1) + (x^2 - 2x + 1) - (2x^2 + 8)$

28. $-(x^3 - 2) + (4x^3 - 2x) - (2x^2 + 3)$

29. $-(5n^2 - 1) - (-3n^2 + 5) - (n^2 - n)$

30. $2(t^2 + 5) - 3(t^2 + 5) + 5(t^2 + 5)$

31. $-10(u + 1) + 8(u - 1) - 3(u + 6)$

32. *Retail Sales* For 1990 through 2000, the total sales (in billions) for retail stores R and for durable-goods stores D can be modeled by

$$R = -0.21t^2 + 31.6t + 357.4$$

$$D = 0.26t^2 + 2.9t + 343$$

where t is the number of years since 1990. Find a model for the sales of non-durable goods stores N. (*Hint:* Retail sales = durable goods sales + non-durable goods sales)

33. *Profit* For 1990 through 2000, the revenue R and cost C of producing a product can be modeled by

$$R = \tfrac{3}{4}t^2 + \tfrac{5}{3}t + 108$$

$$C = \tfrac{7}{12}t^2 + \tfrac{5}{6}t + 94$$

where t is the number of years since 1990. Find a model for the profit P earned from 1990 to 2000. (*Hint:* Profit = Revenue − Cost)

Reteaching with Practice

For use with pages 576–582

GOAL **Add and subtract polynomials and use polynomials to model real-life situations**

VOCABULARY

A **polynomial** is an expression which is the sum of terms of the form ax^k where k is a nonnegative integer.

A polynomial is written in **standard form** when the terms are placed in descending order, from largest degree to smallest degree.

The **degree** of each term of a polynomial is the exponent of the variable.

The **degree of a polynomial** is the largest degree of its terms.

When a polynomial is written in standard form, the coefficient of the first term is the **leading coefficient.**

A **monomial** is a polynomial with only one term.

A **binomial** is a polynomial with two terms.

A **trinomial** is a polynomial with three terms.

EXAMPLE 1 *Adding Polynomials*

_____

Find the sum and write the answer in standard form.

a. $(6x - x^2 + 3) + (4x^2 - x - 2)$

b. $(x^2 - x - 4) + (2x + 3x^2 + 1)$

SOLUTION

a. Vertical format: Write each expression in standard form.
Align like terms.

$$-x^2 + 6x + 3$$
$$\underline{4x^2 - x - 2}$$
$$3x^2 + 5x + 1$$

b. Horizontal format: Add like terms.

$$(x^2 - x - 4) + (2x + 3x^2 + 1) = (x^2 + 3x^2) + (-x + 2x) + (-4 + 1)$$
$$= 4x^2 + x - 3$$

Exercises for Example 1
....

Find the sum.

1. $(7 + 2x - 4x^2) + (-3x + x^2 - 5)$ **2.** $(8x - 9 + 2x^2) + (1 + x - 6x^2)$

Reteaching with Practice

For use with pages 576–582

EXAMPLE 2 *Subtracting Polynomials*

Find the difference and write the answer in standard form.

a. $(5x^2 - 4x + 1) - (8 - x^2)$ **b.** $(-x + 2x^2) - (3x^2 + 7x - 2)$

SOLUTION

a. Vertical format: To subtract, you add the opposite.

$$(5x^2 - 4x + 1)$$
$$\underline{-\qquad (8 - x^2)}$$ Add the opposite.

$$5x^2 - 4x + 1$$
$$\underline{+ \; x^2 \qquad\quad - 8}$$
$$6x^2 - 4x - 7$$

b. Horizontal format:

$$(-x + 2x^2) - (3x^2 + 7x - 2) = -x + 2x^2 - 3x^2 - 7x + 2$$
$$= (2x^2 - 3x^2) + (-x - 7x) + 2$$
$$= -x^2 - 8x + 2$$

Exercises for Example 2

Find the difference.

3. $(x + 7x^2) - (1 + 3x - x^2)$ **4.** $(2x + 3 - 5x^2) - (2x^2 - x + 6)$

EXAMPLE 3 *Using Polynomials in Real-Life*

From 1992 to 1996, the annual sales (in millions of dollars) for
Company D and Company S can be modeled by the following equations,
where t is the number of years since 1992.

Company D model: $D = 316t^2 - 1138t + 3145$

Company S model: $S = 127t^2 - 155t + 3452$

Find a model for the total annual sales A (in millions of dollars) for
Company D and Company S from 1992 to 1996.

SOLUTION

You can find a model for A by adding the models for D and S.

$$316t^2 - 1138t + 3145$$
$$\underline{+ \; 127t^2 - \quad 155t + 3452}$$
$$443t^2 - 1293t + 6597$$

The model for the sum is $A = 443t^2 - 1293t + 6597$.

Exercise for Example 3

5. Find a model for the difference N (in millions of dollars) of
Company D sales and Company S sales from 1992 to 1996.

NAME _____ DATE _____

Quick Catch-Up for Absent Students

For use with pages 575–583

The items checked below were covered in class on (date missed) _____

Activity 10.1: Modeling Addition of Polynomials (p. 575)

_____ **Goal:** Model the addition of polynomials with algebra tiles.

Lesson 10.1: Adding and Subtracting Polynomials

_____ **Goal 1:** Add and subtract polynomials. (pp. 576–577)

Material Covered:

_____ Example 1: Identifying Polynomial Coefficients

_____ Example 2: Classifying Polynomials

_____ Example 3: Adding Polynomials

_____ Student Help: Study Tip

_____ Example 4: Subtracting Polynomials

Vocabulary:

polynomial, p. 576   standard form of a polynomial, p. 576
degree of a term, p. 576   degree of a polynomial, p. 576
leading coefficient, p. 576   monomial, p. 576
binomial, p. 576   trinomial, p. 576

_____ **Goal 2:** Use polynomials to model real-life situations. (pp. 578–579)

Material Covered:

_____ Example 5: Subtracting Polynomials

_____ Example 6: Adding Polynomials

Activity 10.1: Graphing Polynomial Functions (p. 583)

_____ **Goal:** Check an answer when finding the sum or difference of polynomials using a
graphing calculator or computer.

_____ Student Help: Keystroke Help

_____ Other (specify) _____

_____

Homework and Additional Learning Support

_____ Textbook (specify) pp. 579–582 _____

_____

_____ Internet: Extra Examples at www.mcdougallittel.com

_____ *Reteaching with Practice* worksheet (specify exercises)_____

_____ *Personal Student Tutor* for Lesson 10.1

NAME _____ DATE _____

Interdisciplinary Application

For use with pages 576–582

Stained Glass

ART   In art class you are asked to design a stained glass window for a company that is having a new building built. The exact dimensions of the window are not known. The main part of the stained glass window will have a length of $12x + 2$ inches and a height of $6x$ inches and will consist of colors other than blue. Surrounding the main part of the window will be a border x inches wide made out of blue stain glass.

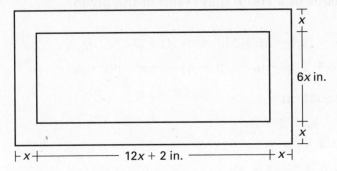

1. Use the verbal model to write a polynomial expression for the amount of blue glass you will use for the border.

$$\boxed{\text{Area of border}} = \boxed{\text{Total Area}} - \boxed{\text{Area of main part}}$$

2. Simplify your polynomial expression in Exercise 1.

3. Use the polynomial expression in Exercise 2 to find the amount of blue glass you would use if the border is 3 inches wide.

4. Use a verbal model to write a polynomial expression for the amount of glass you will use for the main part of the window. Simplify the polynomial expression.

5. Use Guess, Check, and Revise to find the length and height of the main part of the window if its area is 1200 square inches.

6. Find the length and height of the entire stained glass window if the border is 3 inches wide.

7. The company plans to donate $10 per square inch of glass used to your school's general fund. How much money will your school receive from the company?

NAME _____ DATE _____

Challenge: Skills and Applications

For use with pages 576–582

In Exercises 1–3, find the value of *k* that results in the given sum or difference.

1. $[9x^2 - 4x - 8] + [(k - 2)x^2 + 3x + 5] = 5x^2 - x - 3$

2. $(8x^2 - 4kx - 3) - (6x^2 + 3x + 7) = 2x^2 - \frac{7}{2}x - 10$

3. $[(7k - 3)x^2 + 5x + 1] - [(2k - 5)x^2 - 2x + 8] = -x^2 + 7x - 7$

In Exercises 4–5, find the values of *a* and *b* that result in the given sum or difference.

4. $[(a + 1)x^2 + bx + 2] + [(b - 2)x^2 - ax + 5] = x^2 - 4x + 7$

5. $[2ax^2 - (b + 5)x - 4] - [(b + 2)x^2 - (a - 7)x - 8] = -x^2 - 9x + 4$

In Exercises 6–8, solve the equation.

6. $a^2 + 4a - (3a^2 + 2a - 5) = 3a + 9 - 2a^2$

7. $b^2 = (3b^2 + 4b + 7) - (2b^2 + 5b - 2)$

8. $4k - 5 = (k^2 + 5k + 11) + (-k^2 - 4k + 5)$

9. When a certain polynomial is added to $4x^3 - 7x^2 - 9x + 10$, the result is $-3x^3 + x - 4$. What is the polynomial?

10. Suppose $13x^3 - 5x^2 + 4x - 3$ is added to $-8x^2 + 3x + 9$, and suppose $4x^3 + x^2 - 6x - 7$ is added to $3x^2 + 6x + 5$. Find the result when the second of these two sums is subtracted from the first.

11. Find the lateral area (the combined areas of the vertical faces) of the space figure below. Express the area as a simplified polynomial in the variable *x*. (Each angle in the figure is a right angle.)

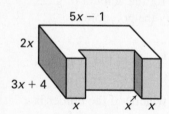

5x − 1
2x
3x + 4
x   x x

Algebra 1
Chapter 10 Resource Book

TEACHER'S NAME _____ CLASS _____ ROOM _____ DATE _____

Lesson Plan

1-day lesson (See *Pacing the Chapter,* TE pages 572C–572D)   **For use with pages 584–589**

 GOALS
1. **Multiply two polynomials.**
2. **Use polynomial multiplication in real-life situations.**

State/Local Objectives _____

_____

✓ Check the items you wish to use for this lesson.

STARTING OPTIONS
_____ Homework Check: TE page 580; Answer Transparencies
_____ Warm-Up or Daily Homework Quiz: TE pages 584 and 582, CRB page 25, or Transparencies

TEACHING OPTIONS
_____ Motivating the Lesson: TE page 585
_____ Lesson Opener (Visual Approach): CRB page 26 or Transparencies
_____ Graphing Calculator Activity with Keystrokes: CRB pages 27–28
_____ Examples 1–5: SE pages 584–586
_____ Extra Examples: TE pages 585–586 or Transparencies; Internet
_____ Closure Question: TE page 586
_____ Guided Practice Exercises: SE page 587

APPLY/HOMEWORK
Homework Assignment
_____ Basic   18–46 even, 54, 55, 60, 65, 70, 73, 75, 80, 85
_____ Average   18–46 even, 54–57, 60, 65, 70, 73, 75, 80, 85
_____ Advanced   18–46 even, 54–57, 60, 61, 65, 70, 73, 75, 80, 85

Reteaching the Lesson
_____ Practice Masters: CRB pages 29–31 (Level A, Level B, Level C)
_____ Reteaching with Practice: CRB pages 32–33 or Practice Workbook with Examples
_____ Personal Student Tutor

Extending the Lesson
_____ Applications (Real-Life): CRB page 35
_____ Challenge: SE page 589; CRB page 36 or Internet

ASSESSMENT OPTIONS
_____ Checkpoint Exercises: TE pages 585–586 or Transparencies
_____ Daily Homework Quiz (10.2): TE page 589, CRB page 39, or Transparencies
_____ Standardized Test Practice: SE page 589; TE page 589; STP Workbook; Transparencies

Notes _____

_____

_____

TEACHER'S NAME _____ CLASS _____ ROOM _____ DATE _____

Lesson Plan for Block Scheduling

Half-day lesson (See *Pacing the Chapter,* TE pages 572C–572D)   For use with pages 584–589

 GOALS   1. **Multiply two polynomials.**
             2. **Use polynomial multiplication in real-life situations.**

State/Local Objectives _____

_____

_____

✓ **Check the items you wish to use for this lesson.**

STARTING OPTIONS

____ Homework Check: TE page 580; Answer Transparencies

____ Warm-Up or Daily Homework Quiz: TE pages 584 and
        582, CRB page 25, or Transparencies

TEACHING OPTIONS

____ Motivating the Lesson: TE page 585

____ Lesson Opener (Visual Approach): CRB page 26 or Transparencies

____ Graphing Calculator Activity with Keystrokes: CRB pages 27–28

____ Examples 1–5: SE pages 584–586

____ Extra Examples: TE pages 585–586 or Transparencies; Internet

____ Closure Question: TE page 586

____ Guided Practice Exercises: SE page 587

APPLY/HOMEWORK

Homework Assignment (See also the assignment for Lesson 10.3.)

____ Block Schedule: 18–46 even, 54–57, 60, 65, 70, 73, 75, 80, 85

Reteaching the Lesson

____ Practice Masters: CRB pages 29–31 (Level A, Level B, Level C)

____ Reteaching with Practice: CRB pages 32–33 or Practice Workbook with Examples

____ Personal Student Tutor

Extending the Lesson

____ Applications (Real-Life): CRB page 35

____ Challenge: SE page 589; CRB page 36 or Internet

ASSESSMENT OPTIONS

____ Checkpoint Exercises: TE pages 585–586 or Transparencies

____ Daily Homework Quiz (10.2): TE page 589, CRB page 39, or Transparencies

____ Standardized Test Practice: SE page 589; TE page 589; STP Workbook; Transparencies

Notes _____

_____

_____

CHAPTER PACING GUIDE	
Day	**Lesson**
1	Assess Ch. 9; 10.1 (all)
2	**10.2 (all)**; 10.3 (begin)
3	10.3 (end); 10.4 (all)
4	10.5 (all)
5	10.6 (all)
6	10.7 (all)
7	10.8 (all)
8	Review/Assess Ch. 10

Lesson 10.2

NAME _____ DATE _____

WARM-UP EXERCISES

For use before Lesson 10.2, pages 584–589

Simplify the expression.

1. $2(x + 4)$

2. $-3(2 - 5x)$

3. $7(12 + 3x) - 10$

4. $(3x)(-4x)$

5. $3x + (-4x)$

DAILY HOMEWORK QUIZ

For use after Lesson 10.1, pages 575–583

Identify the leading coefficient; then classify the polynomial by degree and number of terms.

1. $7 - 3n + 5n^2$

2. $-3x$

Add or subtract.

3. $(4x^2 + 6) + (-x^2 + 3x - 4)$

4. $(x^3 - 2x^2 + 3x - 5) - (x^3 - 3x^2 + 6)$

5. $(b^4 - 2b^3 - b + 5) + (3b^3 + b - 5)$

6. $(8y^2 + 3y) - (6y^2 - 5y - 10)$

Visual Approach Lesson Opener

For use with pages 584–589

Match each area to the correct model. Then use the model to find the product.

1. $x(x)$ **2.** $x(x + 1)$ **3.** $2x(x)$

4. $2x(x + 1)$ **5.** $(x + 2)(x + 1)$ **6.** $(x + 3)(x + 1)$

A.
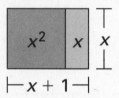

B.
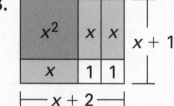

C.
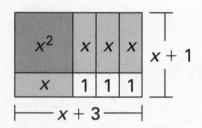

D.
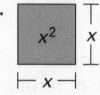

E.
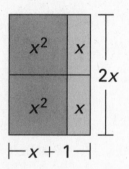

F.
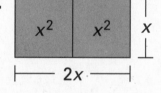

NAME _____ DATE _____

Graphing Calculator Activity

For use with pages 584–589

GOAL **To discover how to multiply two binomials**

You can use the table feature of a graphing calculator to help you
determine if two expressions are equivalent.

X	Y1	Y2
1	21	21
2	40	40
3	63	63
4	90	90
5	121	121
6	156	156

Activity

❶ Enter $(x + 7)(2x + 3)$ into your graphing calculator as equation Y_1.

❷ The three expressions below are possible products for $(x + 7)(2x + 3)$.
To discover which expression is correct, enter each expression into your
graphing calculator as equation Y_2. Then use the *Table* feature to compare
each expression with equation Y_1. When the values of Y_1 and Y_2 in the
table are exactly the same, you have found the correct product.

   a. $2x^2 + 21$ **b.** $3x + 21$ **c.** $2x^2 + 3x + 14x + 21$

❸ Using what you learned in Lesson 10.1, simplify your answer for Step 2.

❹ What property is used twice to multiply binomials?

Exercises

**In Exercises 1–4, use the *Table* feature of your graphing calculator to
determine whether the given product is correct.**

  1. $(x - 4)(x + 9) = x^2 + 5x - 36$ **2.** $(x - 12)(x - 1) = x^2 + 12$

  3. $(8x + 1)(3x - 4) = 11x - 4$ **4.** $(2x - 9)(5x - 6) = 10x^2 - 57x + 54$

**In Exercises 5–10, find the product. Then use the *Table* feature of your
graphing calculator to check your answer.**

  5. $(x + 1)(x + 8)$ **6.** $(x - 10)(x + 9)$ **7.** $(3x - 4)(x - 5)$

  8. $(x + 2)(9x - 2)$ **9.** $(5x + 4)(7x + 3)$ **10.** $(4x - 2)(8x + 9)$

11. Describe in your own words how to multiply binomials.

Lesson 10.2

See page 28 for keystrokes.

Graphing Calculator Activity

For use with pages 584–589

Lesson 10.2

TI-82

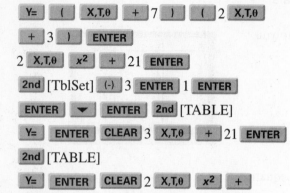

TI-83

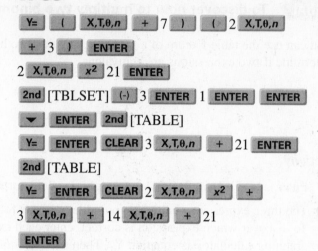

SHARP EL-9600c

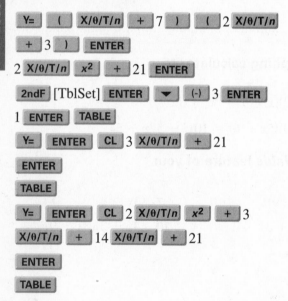

CASIO CFX-9850GA PLUS

From the main menu, choose TABLE.

[ ] [X,θ,T] [+] 7 [ )] [( ] 2 [+] [X,θ,T] [+]

3 [ )] [EXE]

2 [X,θ,T] [x²] [+] 21 [EXE]

[F5] [(-)] 3 [EXE] 3 [EXE] 1 [EXE]

[EXIT] [F6]

[EXIT] [▲] 3 [X,θ,T] [+] 21 [EXE]

[F6]

[EXIT] [▲] 2 [X,θ,T] [x²] [+] 3 [X,θ,T] [+] 14

[X,θ,T] [+] 21 [EXE]

[F6]

NAME _____ DATE _____

Practice A

For use with pages 584–589

Write an equation that represents the product of two binomials as shown in the area model.

1.

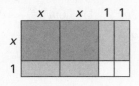

2.

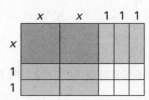

Find the product.

3. $2(3x + 1)$

4. $-4(3x - 5)$

5. $(2x)(5x - 1)$

6. $6n(4 - 5n)$

7. $x^2(3x - 7)$

8. $(8m^2 - 4m + 1)(3m^2)$

9. $(-5t)(t^2 + 2t - 4)$

10. $3x^2(2x^2 - 4x - 7)$

11. $(5a^2 + 3a - 7)(-2a^2)$

Use the distributive property to find the product.

12. $(t + 3)(t + 3)$

13. $(n + 5)(n + 1)$

14. $(2x + 5)(x - 4)$

15. $(4a + 5)(2a - 3)$

16. $(3x^2 + 2x + 1)(x + 3)$

17. $(4x^2 - 3x + 2)(2x + 5)$

Use the FOIL pattern to find the product.

18. $(w + 5)(w + 2)$

19. $(3z + 1)(z + 2)$

20. $(x - 2)(x - 3)$

21. $(4x + 7)(x + 5)$

22. $(2x - 2)(x + 8)$

23. $(5n + 3)(4n - 2)$

Find the product.

24. $(3b - 2)(2b - 3)$

25. $(5x + 4)(3x - 2)$

26. $(10n + 5)(3n - 2)$

27. $(x - 7)(3x + 9)$

28. $(4t + 3)(4t + 3)$

29. $(x^2 + 3x + 1)(x - 2)$

Find an expression for the area of the figure. Give your answer as a quadratic polynomial.

30.

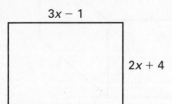

$3x - 1$

$2x + 4$

31.

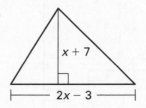

$x + 7$

$2x - 3$

Practice B

For use with pages 584–589

Find the product.

1. $5(2x + 3)$
2. $-7(5x - 3)$
3. $(6x)(3x - 4)$
4. $6x(2x^2 - 5x + 1)$
5. $(4x^2 - 7x)(-x)$
6. $(3m^2 - 1)(-6m^3)$
7. $2x^2(x^3 - x^2 + 8x - 5)$
8. $(6x - 5x^2 + 8)(3x)$
9. $(3a^2 - 7a + 9)(-5a^2)$

Use the distributive property to find the product.

10. $(2x + 3)(x - 1)$
11. $(t + 2)(t + 2)$
12. $(n + 4)(n + 2)$
13. $(3a + 7)(3a - 7)$
14. $(2x^2 - 5x + 3)(x + 4)$
15. $(2x^2 - 5x + 4)(3x + 1)$

Use the FOIL pattern to find the product.

16. $(m + 7)(m + 1)$
17. $(2t + 1)(t + 3)$
18. $(x - 4)(x - 2)$
19. $(3x + 8)(x + 2)$
20. $(5x - 3)(x + 7)$
21. $(6n + 1)(5n - 3)$

Find the product.

22. $(3x + 2)(x + 5)$
23. $(x + 5)(x - 6)$
24. $(x - 8)(x - 4)$
25. $(x - 7)(x + 4)$
26. $(x + 1)(8x - 3)$
27. $(5x - 2)(x - 6)$
28. $(x^2 - 3)(x + 4)$
29. $(x + 5)(x^2 + 4x)$
30. $\left(\frac{1}{2}x + 3\right)(4x + 5)$
31. $\left(\frac{1}{3}x - 2\right)\left(\frac{1}{2}x + 6\right)$
32. $\left(x + \frac{1}{2}\right)\left(2x - \frac{1}{3}\right)$
33. $(3x + 2)(2x + 5)$
34. $(2x - 1)(6x - 7)$
35. $(5x + 2)(5x^2 - 2)$
36. $(4x - 9)(3x + 1)$
37. $(2x^2 + 4)(3x + 1)$
38. $(6x + 5)\left(2x - \frac{1}{3}\right)$
39. $(4x - 1)(8x^2 + 3)$

40. **Floor Plan** The floor plan of a home is shown below. Find an expression for the area of the home. What is the area if $x = 20$ feet?

41. **Volume** Find an expression for the volume of the box. What is the volume if $x = 2$ inches?

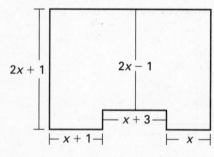

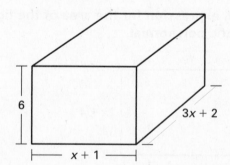

Lesson 10.2

NAME _____ DATE _____

Practice C

For use with pages 584–589

Find the product.

1. $7(3x + 2)$

2. $-9(4x - 6)$

3. $(3x)(5x - 7)$

4. $5x(-2x^2 - 6x + 3)$

5. $(5x^2 - 4x)(-3x)$

6. $\left(4x^2 - \frac{1}{2}\right)(-8x^3)$

7. $5x^2(x^3 - 3x^2 + 2x - 1)$

8. $(6x + 4x^2 - 8)\left(\frac{3}{2}x\right)$

9. $(3x^2 + 6x - 9)\left(-\frac{1}{3}x^2\right)$

Use the distributive property to find the product.

10. $(5x + 2)(x - 4)$

11. $(m + 5)(m - 5)$

12. $(t + 8)(t + 4)$

13. $(4n + 7)(2n - 5)$

14. $(3x^2 - 4x + 1)(x + 5)$

15. $(4x^2 - 6x + 8)\left(2x + \frac{1}{2}\right)$

Use the FOIL pattern to find the product.

16. $(a + 3)(a + 5)$

17. $(2t + 7)(t + 5)$

18. $(x - 6)(x - 4)$

19. $(5x + 1)(3x - 2)$

20. $\left(x - \frac{3}{2}\right)(2x + 4)$

21. $\left(2n + \frac{1}{3}\right)(6n - 3)$

Find the product.

22. $(x - 1)(x + 8)$

23. $(x + 6)(x - 6)$

24. $(x - 8)(x - 4)$

25. $(5x - 2)(3x + 7)$

26. $(x + 8)(4x - 5)$

27. $(7x - 5)(x - 3)$

28. $(x^2 + 4)(x - 5)$

29. $(3x + 4)(x^2 + 5x)$

30. $\left(\frac{1}{3}x - 3\right)(6x + 7)$

31. $\left(\frac{1}{2}x + 8\right)\left(\frac{1}{2}x - 4\right)$

32. $\left(x + \frac{1}{3}\right)\left(x - \frac{1}{6}\right)$

33. $(7x - 2)(2x - 7)$

34. $(2.5x + 1)(3.1x + 2)$

35. $\left(x + \frac{3}{4}\right)\left(x - \frac{1}{4}\right)$

36. $(0.5x - 4)(6x + 2)$

37. $(5x^2 + 3)(2x - 3)$

38. $(7x^2 + 2)\left(5x - \frac{1}{2}\right)$

39. $(6x - 9)(4x^2 + 1)$

Find an expression for the area of the figure. Give your answer as a quadratic polynomial.

40.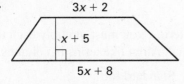
$3x + 2$
$x + 5$
$5x + 8$

41.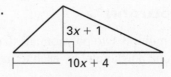
$3x + 1$
$10x + 4$

42. *Exercise Bike* You ride an exercise bike that has an electronic odometer and clock. Each week you increase the rate R and time T at which you ride the bike. The equation $R = \frac{2}{5}x + 14$ models the rate at which you ride, where R is measured in miles per hour and $x = 0$ corresponds to week 0. The equation $T = \frac{1}{30}x + \frac{1}{12}$ models the amount of time you ride at each workout, where T is measured in hours and $x = 0$ corresponds to week 0. Find a model for the distance D you ride in a workout.

Algebra 1
Chapter 10 Resource Book

Reteaching with Practice

For use with pages 584–589

GOAL Multiply two polynomials and use polynomial multiplication in real-life situations.

VOCABULARY

To multiply two binomials, use a pattern called the **FOIL** pattern.
Multiply the First, Outer, Inner, and Last terms.

EXAMPLE 1 *Multiplying Binomials Using the FOIL Pattern*

Find the product $(4x + 3)(x + 2)$.

SOLUTION

$$\overset{\text{F}\quad\text{O}\quad\text{I}\quad\text{L}}{(4x + 3)(x + 2) = 4x^2 + 8x + 3x + 6}$$ Mental math

$$= 4x^2 + 11x + 6$$ Simplify.

Exercises for Example 1

Use the FOIL pattern to find the product.

1. $(2x + 3)(x + 1)$ **2.** $(y - 2)(y - 3)$ **3.** $(3a + 2)(2a - 1)$

EXAMPLE 2 *Multiplying Polynomials Vertically*

Find the product $(x + 3)(4 - 2x^2 + x)$.

SOLUTION

To multiply two polynomials that have three or more terms, you must multiply each term of one polynomial by each term of the other polynomial. Align like terms in columns.

$$
\begin{array}{rl}
-2x^2 + x + 4 & \text{Standard form} \\
\underline{\times \qquad\quad x + 3} & \text{Standard form} \\
-6x^2 + 3x + 12 & 3(-2x^2 + x + 4) \\
\underline{-2x^3 + \quad x^2 + 4x \qquad} & x(-2x^2 + x + 4) \\
-2x^3 - \quad 5x^2 + 7x + 12 & \text{Combine like terms.}
\end{array}
$$

Exercises for Example 2

Multiply the polynomials vertically.

4. $(a + 4)(a^2 + 3 - 2a)$ **5.** $(2y + 1)(y^2 - 5 + y)$

NAME _____ DATE _____

Reteaching with Practice

For use with pages 584–589

EXAMPLE 3 *Multiplying Polynomials Horizontally*

Find the product $(x + 4)(-2x^2 + 3x - 1)$.

SOLUTION

Multiply $-2x^2 + 3x - 1$ by each term of $x + 4$.

$$
\begin{aligned}
(x + 4)(-2x^2 + 3x - 1) &= x(-2x^2 + 3x - 1) + 4(-2x^2 + 3x - 1) \\
&= -2x^3 + 3x^2 - x - 8x^2 + 12x - 4 \\
&= -2x^3 - 5x^2 + 11x - 4
\end{aligned}
$$

Exercises for Example 3

Multiply the polynomials horizontally.

6. $(a + 4)(a^2 + 3 - 2a)$ **7.** $(2y + 1)(y^2 - 5 + y)$

EXAMPLE 4 *Multiplying Binomials to Find an Area*

The dimensions of a rectangular garden can be represented by a width of $(x + 6)$ feet and a length of $(2x + 5)$ feet. Write a polynomial expression for the area A of the garden.

SOLUTION

The area model for a rectangle is $A = (\text{width})(\text{length})$.

$A = (\text{width})(\text{length})$	Area model for a rectangle
$= (x + 6)(2x + 5)$	Substitute $x + 6$ for width and $2x + 5$ for length.
$= 2x^2 + 5x + 12x + 30$	FOIL pattern
$= 2x^2 + 17x + 30$	Combine like terms.

The area A of the garden in square feet can be represented by $2x^2 + 17x + 30$.

Exercise for Example 4

8. Rework Example 4 if the width is $(x + 3)$ feet and the length is $(3x + 2)$ feet.

NAME _____ DATE _____

Quick Catch-Up for Absent Students

For use with pages 584–589

The items checked below were covered in class on (date missed) _____

Lesson 10.2: Multiplying Polynomials

_____ **Goal 1:** Multiply two polynomials. (pp. 584–585)

Material Covered:

_____ Activity: Investigating Binomial Multiplication

_____ Student Help: Look Back

_____ Example 1: Using the Distributive Property

_____ Example 2: Multiplying Binomials Using the FOIL Pattern

_____ Example 3: Multiplying Polynomials Vertically

_____ Student Help: Look Back

_____ Example 4: Multiplying Polynomials Horizontally

Vocabulary:

FOIL pattern, p. 585

_____ **Goal 2:** Use polynomial multiplication in real-life situations. (p. 586)

Material Covered:

_____ Example 5: Multiplying Binomials to Find an Area

_____ Other (specify) _____

_____

Homework and Additional Learning Support

_____ Textbook (specify) _pp. 587–589_____

_____

_____ Internet: Extra Examples at www.mcdougallittel.com

_____ *Reteaching with Practice* worksheet (specify exercises)_____

_____ *Personal Student Tutor* for Lesson 10.2

NAME _____ DATE _____

Real-Life Application: When Will I Ever Use This?

For use with pages 584–589

Cutting the Lawn

Cutting lawns is a popular summer job. It is a great way to earn money and to help others in your community.

Your neighbor asks you to cut her lawn. Her lawn is $(6x + 3)$ feet wide and $(10x + 3)$ feet long. A section of the lawn is fenced in for her dog. The dog cage is $(x - 2)$ feet wide and $(x + 2)$ feet long.

1. Write a polynomial expression that represents the total area of the lawn. Give your answer as a trinomial.

2. Write a polynomial expression that represents the area of the dog cage. Give your answer as a binomial.

3. Write a polynomial expression that represents the area of the lawn that you will mow. Give your answer as a binomial.

4. Copy and complete the table.

x (feet)	6	9	12
Total area of lawn (square feet)			
Area of dog cage (square feet)			
Area of lawn mowed (square feet)			

5. When x is doubled from 6 to 12, do the areas also double? Explain.

Challenge: Skills and Applications

For use with pages 584–589

In Exercises 1–6, find the product.

1. $(2y^3 - 5y)(4y^2 + 3)$

2. $(7 - a^3)(3a^4 - 2)$

3. $(6c^4 - d^2)(2c^4 + d^2)$

4. $2x(x^2 + 8)(3x^2 + 5)$

5. $(5a^2 - 3b^2)(a^2 - 9b^2)$

6. $(x - 3)(x - 4)(x + 2)$

In Exercises 7–9, simplify by multiplying and then adding and subtracting. Write the result as a polynomial in standard form.

7. $(3x + 1)(x + 5) + x^2(x - 4)$

8. $(3x - 6)(3x + 6) + 2x(7 - x^2)$

9. $(x^2 + 3)(x^2 - 5) + x^3(x - 9)$

10. Suppose $(x + a)(x + b) = x^2 + (b - 6)x + 180$. Find the values of a and b.

In Exercises 11–12, solve the equation.

11. $c(3c + 2) - 3(c^2 + 4c - 17) = 1$

12. $6r^2 + 5r = 2(r^2 + 2) - 4(5 - r^2) + 1$

In Exercises 13–15, use the following information.

A jewelry box is made from wood that is $\frac{1}{2}$ inch thick.
The length of the box is $3x$, the width is $2x$, and the height is x.

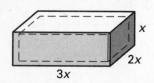

13. The space inside the box (dashed in the diagram) is a rectangular solid. Write an expression for the length, the width, and the height of this space in terms of x.

14. Write a polynomial in standard form that represents the volume of the space inside the box.

15. If the shortest side (x) is 4 inches, what is the volume of the space inside the box?

TEACHER'S NAME _____ CLASS _____ ROOM _____ DATE _____

Lesson Plan

2-day lesson (See *Pacing the Chapter*, TE pages 572C–572D)   **For use with pages 590–596**

GOALS
1. **Use special product patterns for the product of a sum and a difference, and for the square of a binomial.**
2. **Use special products as real-life models.**

State/Local Objectives _____
_____

✓ **Check the items you wish to use for this lesson.**

STARTING OPTIONS

____ Homework Check: TE page 587; Answer Transparencies
____ Warm-Up or Daily Homework Quiz: TE pages 590 and 589, CRB page 39, or Transparencies

TEACHING OPTIONS

____ Motivating the Lesson: TE page 591
____ Lesson Opener (Application): CRB page 40 or Transparencies
____ Examples: Day 1: 1–3, SE page 591; Day 2: 4–5, SE page 592
____ Extra Examples: Day 1: TE page 591 or Transp.; Day 2: TE page 592 or Transp.
____ Closure Question: TE page 592
____ Guided Practice: SE page 593; Day 1: Exs. 1, 4, 6–14; Day 2: Exs. 2–3, 5

APPLY/HOMEWORK

Homework Assignment

____ Basic   Day 1: 15–35; Day 2: 36–42, 47–50, 56–58, 65, 70, 75, 80, 81; Quiz 1: 1–17
____ Average   Day 1: 15–35; Day 2: 36–52, 56–58, 65, 70, 75, 80, 81; Quiz 1: 1–17
____ Advanced   Day 1: 15–35; Day 2: 36–52, 56–62, 65, 70, 75, 80, 81; Quiz 1: 1–17

Reteaching the Lesson

____ Practice Masters: CRB pages 41–43 (Level A, Level B, Level C)
____ Reteaching with Practice: CRB pages 44–45 or Practice Workbook with Examples
____ Personal Student Tutor

Extending the Lesson

____ Cooperative Learning Activity: CRB page 47
____ Applications (Interdisciplinary): CRB page 48
____ Math & History: SE page 596; CRB page 49; Internet
____ Challenge: SE page 595; CRB page 50 or Internet

ASSESSMENT OPTIONS

____ Checkpoint Exercises: Day 1: TE page 591 or Transp.; Day 2: TE page 592 or Transp.
____ Daily Homework Quiz (10.3): TE page 595, CRB page 54, or Transparencies
____ Standardized Test Practice: SE page 595; TE page 595; STP Workbook; Transparencies
____ Quiz (10.1–10.3): SE page 596; CRB page 51

Notes _____
_____
_____

Lesson Plan for Block Scheduling

1-day lesson (See *Pacing the Chapter,* TE pages 572C–572D)   For use with pages 590–596

GOALS 1. **Use special product patterns for the product of a sum and a difference, and for the square of a binomial.**
2. **Use special products as real-life models.**

State/Local Objectives _____

_____

_____

CHAPTER PACING GUIDE	
Day	**Lesson**
1	Assess Ch. 9; 10.1 (all)
2	10.2 (all); **10.3 (begin)**
3	**10.3 (end)**; 10.4 (all)
4	10.5 (all)
5	10.6 (all)
6	10.7 (all)
7	10.8 (all)
8	Review/Assess Ch. 10

✓ **Check the items you wish to use for this lesson.**

STARTING OPTIONS

____ Homework Check: TE page 587; Answer Transparencies
____ Warm-Up or Daily Homework Quiz: TE pages 590 and
          589, CRB page 39, or Transparencies

TEACHING OPTIONS

____ Motivating the Lesson: TE page 591
____ Lesson Opener (Application): CRB page 40 or Transparencies
____ Examples: Day 2: 1–3, SE page 591; Day 3: 4–5, SE page 592
____ Extra Examples: Day 2: TE page 591 or Transp.; Day 3: TE page 592 or Transp.
____ Closure Question: TE page 592
____ Guided Practice: SE page 593   Day 2: Exs. 1, 4, 6–14; Day 3: Exs. 2–3, 5

APPLY/HOMEWORK

Homework Assignment (See also the assignments for Lessons 10.2 and 10.4.)

____ Block Schedule: Day 2: 15–35; Day 3: 36–52, 56–58, 65, 70, 75, 80, 81; Quiz 1: 1–17

Reteaching the Lesson

____ Practice Masters: CRB pages 41–43 (Level A, Level B, Level C)
____ Reteaching with Practice: CRB pages 44–45 or Practice Workbook with Examples
____ Personal Student Tutor

Extending the Lesson

____ Cooperative Learning Activity: CRB page 47
____ Applications (Interdisciplinary): CRB page 48
____ Math & History: SE page 596; CRB page 49; Internet
____ Challenge: SE page 595; CRB page 50 or Internet

ASSESSMENT OPTIONS

____ Checkpoint Exercises: Day 2: TE page 591 or Transp.; Day 3: TE page 592 or Transp.
____ Daily Homework Quiz (10.3): TE page 595, CRB page 54, or Transparencies
____ Standardized Test Practice: SE page 595; TE page 595; STP Workbook; Transparencies
____ Quiz (10.1–10.3): SE page 596; CRB page 51

Notes _____

_____

_____

NAME _____ DATE _____

WARM-UP EXERCISES

For use before Lesson 10.3, pages 590–596

Simplify the expression.

1. $(2x)^2$

2. $(-8m)^2$

3. $\left(\dfrac{1}{4}y\right)^2$

4. $(b^3)^2$

5. $(3n^2)^4$

....

DAILY HOMEWORK QUIZ

For use after Lesson 10.2, pages 584–589

Find the product.

1. $(2x)(8 - x - 3x^2)$

2. $(3t + 5)(t - 3)$

3. $(y - 4)(y + 4)$

4. $(2x + 10)\left(\dfrac{1}{2}x + 2\right)$

5. Write a polynomial expression for the area of the rectangle.

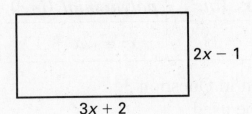

$2x - 1$

$3x + 2$

Algebra 1
Chapter 10 Resource Book

NAME _____ DATE _____

Application Lesson Opener

For use with pages 590–596

In Questions 1–4, use the following information.

Uri is building a house. In the original plans for the house, all of the square windows are the same size. Let x represent the length in inches of a side of each window.

1. Uri started the table below to show the area of a window if each side is increased by the same amount. Complete the table. Use FOIL for the last column.

Amount of increase	Length of side (in.)	New area as product of two factors (in.²)	New area as polynomial (in.²)
1 inch	$x + 1$	$(x + 1)(x + 1)$	$x^2 + 2x + 1$
2 inches	$x + 2$	$(x + 2)(x + 2)$	
3 inches	$x + 3$		
4 inches			

2. You used FOIL to fill in the last column in Question 1. Suggest a possible shortcut that could be used.

3. Make a table like the one above to show the area of a window if each dimension is decreased by the same amount. The first row is completed for you. Use FOIL to fill in the last column.

Amount of increase	Length of side (in.)	New area as product of two factors (in.²)	New area as polynomial (in.²)
1 inch	$x - 1$	$(x - 1)(x - 1)$	$x^2 - 2x + 1$

4. You used FOIL to fill in the last column in Question 3. Suggest a possible shortcut that could be used.

Algebra 1
Chapter 10 Resource Book

Practice A

For use with pages 590–596

Find the missing term.

1. $(x - y)^2 = x^2 - $ _____ $ + y^2$

2. $(a + b)^2 = a^2 + $ _____ $ + b^2$

3. $(m + n)^2 = m^2 + $ _____ $ + n^2$

4. $(2x - 5)^2 = 4x^2 - $ _____ $ + 25$

5. $(x - y)(x + y) = x^2 - $ _____

6. $(a + 2b)(a - 2b) = a^2 - $ _____

Write the product of the sum and difference.

7. $(x + 2)(x - 2)$

8. $(t + 3)(t - 3)$

9. $(x + 9)(x - 9)$

10. $(5 + c)(5 - c)$

11. $(n + 5)(n - 5)$

12. $(2x + 7)(2x - 7)$

13. $(7 + d)(7 - d)$

14. $(3x + 1)(3x - 1)$

15. $(5x + 3)(5x - 3)$

Write the square of the binomial as a trinomial.

16. $(x + 4)^2$

17. $(x - 5)^2$

18. $(x + 8)^2$

19. $(2t + 3)^2$

20. $(3y - 5)^2$

21. $(4m - 3)^2$

22. $(2m + 4)^2$

23. $(2y + 9)^2$

24. $(2k - 3)^2$

Find the product.

25. $(w + 5)(w - 5)$

26. $(3z + 1)(3z - 1)$

27. $(x - 2)(x + 2)$

28. $(4x + 3)(4x - 3)$

29. $(2x - 9)(2x + 9)$

30. $(5n + 1)^2$

31. $(x - 5)^2$

32. $(3x - 2)^2$

33. $(7b + 3)^2$

Use mental math to find the product.

34. $18 \cdot 22$

35. $27 \cdot 33$

36. $54 \cdot 46$

37. *Blue Eyes–Brown Eyes* In humans, the brown eye gene B is dominant and the blue eye gene b is recessive. This means that humans whose eye genes are BB, Bb, or bB have brown eyes and those with bb have blue eyes. The Punnett square at the right shows the results of eye colors for children of parents who each have one B gene and one b gene. What percentage of children will have brown eyes? What percentage will have blue eyes? Use the model $(0.5B + 0.5b)^2 = 0.25BB + 0.5Bb + 0.25bb$ to answer the question.

	B	b
B	BB	Bb
b	bB	bb

Practice B

For use with pages 590–596

Find the missing term.

1. $(m - n)^2 = m^2 - $ _____ $+ n^2$

2. $(x + y)^2 = x^2 + $ _____ $+ y^2$

3. $(2a + 3b)^2 = 4a^2 + $ _____ $+ 9b^2$

4. $(-5x - 2)^2 = 25x^2 + $ _____ $+ 4$

5. $(2w - 7)(2w + 7) = 4w^2 - $ _____

6. $(9 + 2c)(9 - 2c) = 81 - $ _____

Write the product of the sum and difference.

7. $(x + 3)(x - 3)$

8. $(t + 7)(t - 7)$

9. $(2x + 1)(2x - 1)$

10. $(4x + 3)(4x - 3)$

11. $(3n + 3)(3n - 3)$

12. $(5x + 2)(5x - 2)$

13. $(3 + 2d)(3 - 2d)$

14. $(7x + 5)(7x - 5)$

15. $(x + y)(x - y)$

16. $(5x + y)(5x - y)$

17. $(x - 4y)(x + 4y)$

18. $(2x + 3y)(2x - 3y)$

Write the square of the binomial as a trinomial.

19. $(x + 5)^2$

20. $(x - 6)^2$

21. $(x + 9)^2$

22. $(2t + 1)^2$

23. $(4y - 1)^2$

24. $(m + 7)^2$

25. $(m - 2)^2$

26. $(3y - 4)^2$

27. $(3k + 8)^2$

28. $(x - 3)^2$

29. $(5t - 2)^2$

30. $(4n + 5)^2$

Find the product.

31. $(k + 7)(k - 7)$

32. $(3m + 5)(3m - 5)$

33. $(p - q)(p + q)$

34. $\left(\frac{1}{2}x + 4\right)\left(\frac{1}{2}x - 4\right)$

35. $(9x - 7)(9x + 7)$

36. $(6n + 3)^2$

37. $(y - x)^2$

38. $(5b - 7)^2$

39. $(7x + 1)^2$

Use mental math to find the product.

40. $37 \cdot 43$

41. $45 \cdot 35$

42. $82 \cdot 78$

43. *Total Profit* For 1990 through 2000, the number of units N produced by a manufacturing plant can be modeled by $N = 2t + 3$, where N is measured in thousands of units and t is the number of years since 1990. The profit per unit P can be modeled by $P = 2t - 3$, where P is measured in dollars and t is the number of years since 1990. Find an expression for the total profit T, where T is measured in thousands of dollars. What was the total profit in 1995?

Lesson 10.3

Practice C

For use with pages 590–596

Write the product of the sum and difference.

1. $(x + 5)(x - 5)$

2. $(t + 4)(t - 4)$

3. $(3x + 5)(3x - 5)$

4. $(7x + 6)(7x - 6)$

5. $(5n + 5)(5n - 5)$

6. $(9x + 3)(9x - 3)$

7. $(6 + 4d)(6 - 4d)$

8. $\left(8x + \frac{5}{2}\right)\left(8x - \frac{5}{2}\right)$

9. $(m + n)(m - n)$

10. $\left(\frac{1}{3}x + y\right)\left(\frac{1}{3}x - y\right)$

11. $\left(x - \frac{4}{5}y\right)\left(x + \frac{4}{5}y\right)$

12. $(6x + 5y)(6x - 5y)$

Write the square of the binomial as a trinomial.

13. $(x + 3)^2$

14. $(x - 9)^2$

15. $(x + 2y)^2$

16. $(2m + 3)^2$

17. $(7y - 3)^2$

18. $(5y - 3)^2$

19. $\left(b - \frac{2}{3}\right)^2$

20. $\left(m + \frac{1}{2}\right)^2$

21. $(8k + 3)^2$

22. $(x - 0.3)^2$

23. $(7c - 2d)^2$

24. $(5n + 4m)^2$

Find the product.

25. $(c + 5)(c - 5)$

26. $(5m + 2)(5m - 2)$

27. $(v - w)(v + w)$

28. $\left(\frac{1}{3}x + 6\right)\left(\frac{1}{3}x - 6\right)$

29. $(9x - 7)(9x + 7)$

30. $(8n + 5)^2$

31. $(3y - 2x)^2$

32. $(5b - 7c)^2$

33. $(7x + y)^2$

Use mental math to find the product.

34. $36 \cdot 44$

35. 18^2

36. 52^2

37. *Area Model* Find an expression for the area of the shaded region shown below. Then evaluate the expression when x is equal to 5 inches; to 6 inches; and to 7 inches.

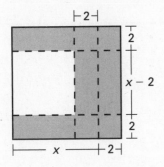

38. *Area Model* Find an expression for the area of the shaded region shown below. Then evaluate the expression when x is equal to 5 inches; to 7 inches; and to 9 inches.

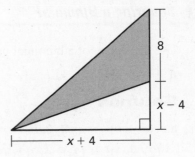

Reteaching with Practice

For use with pages 590–596

GOAL Use special product patterns for the product of a sum and a difference and for the square of a binomial and use special products in real-life models

---

VOCABULARY

Some pairs of binomials have **special product** patterns as follows.

Sum and Difference Pattern

$$(a + b)(a - b) = a^2 - b^2$$

Square of a Binomial Pattern

$$(a + b)^2 = a^2 + 2ab + b^2$$

$$(a - b)^2 = a^2 - 2ab + b^2$$

EXAMPLE 1 *Using the Sum and Difference Pattern*

Use the sum and difference pattern to find the product $(4y + 3)(4y - 3)$.

SOLUTION

$(a + b)(a - b) = a^2 - b^2$	Write pattern.
$(4y + 3)(4y - 3) = (4y)^2 - 3^2$	Apply pattern.
$= 16y^2 - 9$	Simplify.

Exercises for Example 1

Use the sum and difference pattern to find the product.

1. $(x + 5)(x - 5)$ **2.** $(3x + 2)(3x - 2)$ **3.** $(x + 2y)(x - 2y)$

EXAMPLE 2 *Squaring a Binomial*

Use the square of a binomial pattern to find the product.

a. $(2x + 3)^2$ **b.** $(4x - 1)^2$

SOLUTION

a.

$(a + b)^2 = a^2 + 2ab + b^2$	Write pattern.
$(2x + 3)^2 = (2x)^2 + 2(2x)(3) + 3^2$	Apply pattern.
$= 4x^2 + 12x + 9$	Simplify.

b.

$(a - b)^2 = a^2 - 2ab + b^2$	Write pattern.
$(4x - 1)^2 = (4x)^2 - 2(4x)(1) + 1^2$	Apply pattern.
$= 16x^2 - 8x + 1$	Simplify.

NAME _____ DATE _____

Reteaching with Practice

For use with pages 590–596

Exercises for Example 2

Use the square of a binomial pattern to find the product.

4. $(m + n)^2$ **5.** $(3x - 2)^2$ **6.** $(7y + 2)^2$

EXAMPLE 3 **Applying a Special Product Pattern to Find an Area**

Use a special product pattern to find an expression for the area of the shaded region.

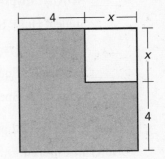

SOLUTION

Verbal Model	Area of shaded region	=

	Area of entire square	−	Area of smaller square

Labels   Area of shaded region = A (square units)

           Area of entire square = $(x + 4)^2$ (square units)

           Area of smaller square = x^2 (square units)

Algebraic $A = (x + 4)^2 - x^2$ Write algebraic model.

Model $= (x^2 + 8x + 16) - x^2$ Apply pattern.

           $= 8x + 16$ Simplify.

The area of the shaded region can be represented by $8x + 16$ square units.

Exercises for Example 3

7. Use a special product pattern to find an expression for the area of the shaded region.

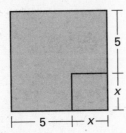

8. Use a special product pattern to find an expression for the area of the shaded region.

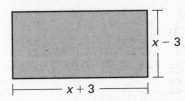

NAME _____ DATE _____

Quick Catch-Up for Absent Students

For use with pages 590–596

The items checked below were covered in class on (date missed) _____

Lesson 10.3: Special Products of Polynomials

_____ **Goal 1:** Use special product patterns for the product of a sum and a difference, and for the square of a binomial. (pp. 590–591)

Material Covered:

_____ Activity: Investigating Special Product Patterns

_____ Student Help: Study Tip

_____ Example 1: Using the Sum and Difference Pattern

_____ Example 2: Squaring a Binomial

_____ Example 3: Special Products and Mental Math

_____ **Goal 2:** Use special products as real-life models. (p. 592)

Material Covered:

_____ Example 4: Finding an Area

_____ Example 5: Modeling a Punnett Square

_____ Other (specify) _____

_____

Homework and Additional Learning Support

_____ Textbook (specify) pp. 593–596 _____

_____

_____ *Reteaching with Practice* worksheet (specify exercises) _____

_____ *Personal Student Tutor* for Lesson 10.3

NAME _____ DATE _____

Cooperative Learning Activity

For use with pages 590–596

GOAL **To use geometry to explain multiplication of polynomials**

Materials: paper, pencil, scissors, glue

Exploring Polynomial Multiplication

Geometry provides a visual way to explain algebraic relationships. In this activity, you and your partner will use two squares to explain polynomial multiplication. The side of the large square has a length of a, and the side of the small square has a length of b.

Instructions

❶ Copy the squares at the left on a piece of unlined paper. Make several copies.

❷ Cut and paste the squares to make a square with sides of length $a + b$. Then create a square with sides of length $a - b$. Label each square accordingly.

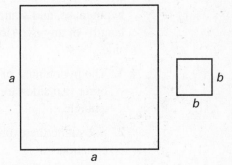

Analyzing the Results

1. What is the area of each of the original squares?

2. What is the area of each of the squares you created?

3. Use the squares to explain the solution of the following equation.

$$2(a^2 + b^2) = \underline{\hspace{3cm}}$$

Interdisciplinary Application

For use with pages 590–596

Pythagoras

GEOMETRY   Little is known of Pythagoras' early life, but
scholars believe that he was born on the island of Samos. In
about 529 B.C., he settled in Crotona, Italy. There he founded
a school (brotherhood) among the aristocrats of the city.
Pythagoras was a Greek philosopher and mathematician. He
became famous for formulating the *Pythagorean theorem*,
even though its principles were known earlier. For example,
the ancient Egyptians had used these principles to create
fields with square corners.

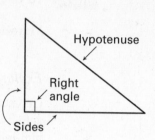

The theorem states that the square of the hypotenuse of a right triangle is equal
to the sum of the squares of the other two sides. The Pythagorean theorem
written as a formula is $c^2 = a^2 + b^2$. In this formula, c is the length of the
hypotenuse, and a and b are the lengths of the other two sides. If you know the
lengths of any two sides of a right triangle, you can find the length of the third
side.

1. The hypotenuse of a right triangle has a length of $t + 11$. The lengths of the
   other two sides are $t + 10$ and $t - 7$. Draw a diagram to model this right
   triangle.

2. Substitute the expressions in the Pythagorean Theorem. Simplify your
   equation.

3. Using your simplified equation from Exercise 2, check the values of $t = 14$
   and $t = 2$.

4. Using the value(s) of t that make the equation true in Exercise 3, find the
   lengths of the three sides of the right triangle. Explain your results.

In Exercises 5–8, use the following information.

A pine tree casts a shadow. The height of the tree in feet is $t - 30$. You stand at
the end of the shadow. The distance between you and the top of the tree in feet
is $t + 20$. The length of the shadow in feet is $t + 19$.

5. Draw a diagram of the pine tree with its shadow. Label the lengths of your
   diagram.

6. Substitute the expressions in the Pythagorean theorem. Simplify your
   equation.

7. Using your simplified equation from Exercise 5, check the values of $t = 21$
   and $t = 41$.

8. Using the value(s) of t that make the equation true in Exercise 7, find the
   length of the tree's shadow, the height of the pine tree, and the distance
   between the ground where you are standing and the top of the tree. Explain
   your results.

Algebra 1
Chapter 10 Resource Book

NAME _____ DATE _____

Math and History Application

For use with page 596

HISTORY   Gregor Mendel was born in Heinzendorf, Austria, in 1822. His parents were farmers and did not have enough money to pay for Gregor's education at the Gymnasium (which is similar to high school). To earn money, Mendel tutored students. Mendel was an excellent student. Upon his graduation from the Gymnasium, Mendel's father wanted him to return home and take over the family farm. Mendel, however, wanted to continue his education. He enrolled in the University of Olmutz for two years, where he studied mathematics and philosophy. Again, Mendel did not have enough money to continue his education. He decided to join the monastery at Brno, which allowed Mendel to continue his education. He also began teaching mathematics and other classes at a local school.

MATH   Mendel began researching heredity in 1856. He grew pea plants in the monastery gardens. He tested plants for two years to be sure traits were constant. It is estimated that Mendel grew as many as 28,000 pea plants during his experiments. In his experiments with heredity, he crossed pea plants with round seeds and pea plants with wrinkled seeds. The results were about 75% of the offspring had round seeds and about 25% of the offspring had wrinkled seeds. From these results and other experiments, Mendel was able to explain basic principles of heredity.

Mendel published his results in a scientific journal and sent 40 copies of the results to scientists and universities. His results went unnoticed until 1900. At that time three other scientists found similar results separately but then realized that Mendel had already published his results in 1866. Gregor Mendel is now given credit for his work in the field of genetics.

1. Results of Mendel's pea plant experiment with a cross of plants with round seeds and wrinkled seeds were 5474 plants with round seeds and 1850 plants with wrinkled seeds. What is the probability that a pea plant will have wrinkled seeds? What is the probability that a pea plant will have round seeds? What is the ratio between round and wrinkled seeds?

2. Mendel also conducted experiments with pea plants that had long stems and short stems. When these plants were crossed, the results were 787 long-stem plants and 277 short-stem plants. What is the ratio between long-stem pea plants and short-stem pea plants?

3. Compare the ratios from Exercises 1 and 2 to the results Mendel found. How do the ratios compare?

NAME _____ DATE _____

Challenge: Skills and Applications

For use with pages 590–596

In Exercises 1–4, simplify by multiplying and then adding and subtracting. Write the result as a polynomial in standard form.

1. $(x - 3)^2 + (x + 5)^2$

2. $(x - 4)(x + 4) - (x + 2)^2$

3. $(2x - 1)^2 - (x + 3)^2$

4. $(2x + 9)(3x - 7) - (5x + 2)^2$

5. Use the distributive property to show that

$$(a + b + c)^2 = a^2 + b^2 + c^2 + 2ab + 2ac + 2bc.$$

In Exercises 6–7, use the results from Exercise 5 to find the product.

6. $(x - 3y + 2z)^2$

7. $(x^2 + 4x - 5)^2$

8. Find the product.

  a. $(x - 5)(x^2 + 5x + 25)$

  b. $(x - 2)(x^2 + 2x + 4)$

  c. Use the result of parts (a) and (b) to suggest a general product formula involving $(x - a)$ and another factor. Use multiplication to test your formula.

9. a. Simplify $(x - 5)(x^3 + 5x^2 + 25x + 125)$.

  b. From your answer to part (a), suggest another product formula like the one from Exercise 8.

In Exercises 10–12, Rick claims that if you multiply four consecutive integers together and add 1, you always get a perfect square.

10. Show that Rick's statement is true if the smallest of the integers is 2.

11. Show that Rick's statement is true if the smallest of the integers is 3.

12. Suppose the smallest of the four integers is $n - 1$. Find a polynomial expression in terms of the variable n that represents the number Rick is talking about. Give the expression in standard form.

13. Show that the expression you wrote in Exercise 12 is the square of the trinomial $n^2 + n - 1$.

Algebra 1
Chapter 10 Resource Book

NAME _____ DATE _____

Quiz 1

For use after Lessons 10.1–10.3

1. Add the polynomials. *(Lesson 10.1)*

$$(3n - 4n^2 + 7) + (7n^2 + 5n - 6)$$

2. Subtract the polynomials. *(Lesson 10.1)*

$$(7y^3 + 4y^2 - 5y + 3) - (8y^3 - 4y^2 - 2y + 2)$$

3. Find the product of the polynomials. *(Lesson 10.2)*

$$-3t^2(3t^2 - 4t + 5)$$

4. Use the FOIL pattern to multiply $(4y + 3)(2y - 2)$. *(Lesson 10.2)*

5. Write $(m - 4)^2$ as a trinomial. *(Lesson 10.3)*

6. Multiply $(3a - 5)(3a + 5)$. *(Lesson 10.3)*

7. You have a rectangular vegetable garden that is 10 feet long and 12 feet wide. You enlarge the garden by increasing each side by y feet. Write the area of the larger garden in terms of a trinomial. *(Lesson 10.3)*

Answers

1. _____

2. _____

3. _____

4. _____

5. _____

6. _____

7. _____

LESSON
10.4

TEACHER'S NAME _____ CLASS _____ ROOM _____ DATE _____

Lesson Plan

1-day lesson (See *Pacing the Chapter,* TE pages 572C–572D)   **For use with pages 597–602**

GOALS 1. **Solve a polynomial equation in factored form.**
2. **Relate factors and *x*-intercepts.**

State/Local Objectives _____

_____

✓ **Check the items you wish to use for this lesson.**

STARTING OPTIONS
____ Homework Check: TE page 593; Answer Transparencies
____ Warm-Up or Daily Homework Quiz: TE pages 597 and 595, CRB page 54, or Transparencies

TEACHING OPTIONS
____ Motivating the Lesson: TE page 598
____ Lesson Opener (Application): CRB page 55 or Transparencies
____ Examples 1–5: SE pages 598–599
____ Extra Examples: TE pages 598–599 or Transparencies
____ Closure Question: TE page 599
____ Guided Practice Exercises: SE page 600

APPLY/HOMEWORK
Homework Assignment
____ Basic   20–42 even, 44–54, 59, 65, 70, 75, 80–82
____ Average   20–42 even, 44–56, 59, 65, 70, 75, 80–82
____ Advanced   20–42 even, 44–56, 59, 60, 65, 70, 75, 80–82

Reteaching the Lesson
____ Practice Masters: CRB pages 56–58 (Level A, Level B, Level C)
____ Reteaching with Practice: CRB pages 59–60 or Practice Workbook with Examples
____ Personal Student Tutor

Extending the Lesson
____ Applications (Real-Life): CRB page 62
____ Challenge: SE page 602; CRB page 63 or Internet

ASSESSMENT OPTIONS
____ Checkpoint Exercises: TE pages 598–599 or Transparencies
____ Daily Homework Quiz (10.4): TE page 602, CRB page 66, or Transparencies
____ Standardized Test Practice: SE page 602; TE page 602; STP Workbook; Transparencies

Notes _____

_____

_____

Lesson 10.4

52   **Algebra 1**
Chapter 10 Resource Book

Copyright © McDougal Littell Inc.
All rights reserved.

TEACHER'S NAME _____ CLASS _____ ROOM _____ DATE _____

Lesson Plan for Block Scheduling

Half-day lesson (See *Pacing the Chapter,* TE pages 572C–572D) For use with pages 597–602

GOALS 1. **Solve a polynomial equation in factored form.**
2. **Relate factors and *x*-intercepts.**

State/Local Objectives _____

_____

_____

✓ **Check the items you wish to use for this lesson.**

STARTING OPTIONS

____ Homework Check: TE page 593; Answer Transparencies
____ Warm-Up or Daily Homework Quiz: TE pages 597 and
      595, CRB page 54, or Transparencies

TEACHING OPTIONS

____ Motivating the Lesson: TE page 598
____ Lesson Opener (Application): CRB page 55 or Transparencies
____ Examples 1–5: SE pages 598–599
____ Extra Examples: TE pages 598–599 or Transparencies
____ Closure Question: TE page 599
____ Guided Practice Exercises: SE page 600

APPLY/HOMEWORK

Homework Assignment (See also the assignment for Lesson 10.3.)

____ Block Schedule: 20–42 even, 44–56, 59, 65, 70, 75, 80–82

Reteaching the Lesson

____ Practice Masters: CRB pages 56–58 (Level A, Level B, Level C)
____ Reteaching with Practice: CRB pages 59–60 or Practice Workbook with Examples
____ Personal Student Tutor

Extending the Lesson

____ Applications (Real-Life): CRB page 62
____ Challenge: SE page 602; CRB page 63 or Internet

ASSESSMENT OPTIONS

____ Checkpoint Exercises: TE pages 598–599 or Transparencies
____ Daily Homework Quiz (10.4): TE page 602, CRB page 66, or Transparencies
____ Standardized Test Practice: SE page 602; TE page 602; STP Workbook; Transparencies

Notes _____

_____

_____

| CHAPTER PACING GUIDE ||
Day	Lesson
1	Assess Ch. 9; 10.1 (all)
2	10.2 (all); 10.3 (begin)
3	10.3 (end); **10.4 (all)**
4	10.5 (all)
5	10.6 (all)
6	10.7 (all)
7	10.8 (all)
8	Review/Assess Ch. 10

NAME _____ DATE _____

WARM-UP EXERCISES

For use before Lesson 10.4, pages 597–602

Solve the equation.

1. $x^2 = 9$

2. $x^2 - 13 = 12$

3. $\dfrac{1}{4}x^2 = 16$

Write each product.

4. $(x - 3)(2x - 1)$

5. $(x + 8)^2$

DAILY HOMEWORK QUIZ

For use after Lesson 10.3, pages 590–596

Find the product.

1. $(5x + 5)(5x - 5)$

2. $(x - 11)^2$

3. $(3n + 4m)^2$

4. The Punnett square shows the possible outcomes when 2 coins are tossed. The outcome on each coin can be modeled by $\dfrac{1}{2}H + \dfrac{1}{2}T$. Show how the square of a binomial can be used to model the results when 2 coins are tossed.

	H	T
H	HH	HT
T	TH	TT

Lesson 10.4

Application Lesson Opener

For use with pages 597–602

In Questions 1–4, use the following information.

You have found that you can sell about 100 candles at $2 each at craft fairs. You decide to increase the price of each candle. You estimate that for each increase of $.10, you will lose 1 customer. Let x represent the number of $.10 price increases. The expression $(2 + 0.1x)(100 - x)$ models the amount you will collect at each fair with each $.10 price increase.

1. You know that if you raise the price too many times, no one will buy your candles. Explain why the equation $(2 + 0.1x)(100 - x) = 0$ describes this situation.

2. If $2 + 0.1x = 0$ will the equation be true? Explain.

3. If $100 - x = 0$, will the equation be true? Explain.

4. What do you think the solutions to the equation are? Use your answers to Questions 2 and 3 to justify your answer.

In Questions 5–8, use the following information.

You also sell about 50 holders at $4 each. You decide to increase the price of each holder. You estimate that for each increase of $.50, you will lose 2 customers. Let y represent the number of $.50 price increases. The expression $(4 + 0.50y)(50 - 2y)$ models the amount you will collect at each fair with each $2 price increase.

5. The equation $(4 + 0.50y)(50 - 2y) = 0$ describes the situation in which no one buys a holder. Why?

6. If $4 + 0.50y = 0$, will the equation be true? Explain.

7. If $50 - 2y = 0$, will the equation be true? Explain.

8. What do you think the solutions to the equation are? Use your answers to Questions 6 and 7 to justify your answer.

Lesson 10.4

NAME _____ DATE _____

Practice A

For use with pages 597–602

Decide if the graph of the function has *x*-intercepts of 2 and −3.

1. $y = (x + 2)(x + 3)$
2. $y = (x - 2)(x - 3)$
3. $y = (x - 2)(x + 3)$
4. $y = 3(x - 2)(x + 3)$
5. $y = -(x + 2)(x - 3)$
6. $y = -4(x - 2)(x + 3)$

Use the zero-product property to solve the equation.

7. $(x + 2)(x + 5) = 0$
8. $(t + 3)(t - 3) = 0$
9. $(x + 1)(x - 5) = 0$

10. $(c + 5)(c + 3) = 0$
11. $(n - 3)(n - 5) = 0$
12. $(x + 7)(x - 7) = 0$

13. $(d + 4)(d + 8) = 0$
14. $(x + 1)(x - 1) = 0$
15. $(x + 5)(x - 3) = 0$

16. $\left(m + \frac{1}{4}\right)\left(m + \frac{3}{8}\right) = 0$
17. $(w + 1.2)(w - 7.1) = 0$
18. $(y + 5)^2 = 0$

Solve the equation.

19. $(2x + 6)(x + 5) = 0$
20. $(3y - 1)(2y + 8) = 0$
21. $(4m + 12)(3m - 15) = 0$

22. $(2n - 8)(5n + 25) = 0$
23. $(2x - 7)(9x - 18) = 0$
24. $(7t + 14)(t + 5) = 0$

25. $(3x + 4)(7x - 21) = 0$
26. $(2c + 6)(3c + 6) = 0$
27. $(2w - 1)(4w - 3) = 0$

28. $(4x + 3)(4x - 3) = 0$
29. $(2x - 9)(2x + 8)^2 = 0$
30. $(5n + 1)(3n - 6) = 0$

Match the function with its graph.

A. $y = (x + 1)(x - 5)$
B. $y = (x - 1)(x + 5)$
C. $y = (x - 1)(x - 5)$

31.
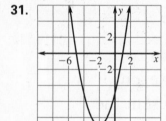

32.
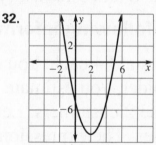

33.
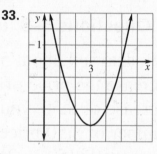

34. An object is dropped from a hot-air balloon 1600 feet above the ground. The height of the object is given by

$$h = -16(t - 10)(t + 10)$$

where the height *h* is measured in feet, and the time *t* is measured in seconds. When will the object hit the ground? Can you see a quick way to find the answer? Explain.

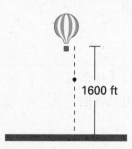

1600 ft

Practice B

For use with pages 597–602

Use the zero-product property to solve the equation.

1. $(x + 1)(x + 6) = 0$

2. $(t + 4)(t - 4) = 0$

3. $(x + 9)(x - 8) = 0$

4. $(c + 7)(c + 2) = 0$

5. $(n - 8)(n - 9) = 0$

6. $(x + 4.2)(x - 4.2) = 0$

7. $\left(d + \frac{3}{4}\right)\left(d + \frac{5}{8}\right) = 0$

8. $\left(x + \frac{1}{2}\right)\left(x - \frac{1}{2}\right) = 0$

9. $(x + 5.4)(x - 3) = 0$

10. $5(m + 4)^2 = 0$

11. $(x - 3.2)\left(w - \frac{3}{2}\right) = 0$

12. $(y - 6)(y + 6)^2 = 0$

Solve the equation.

13. $(2x + 8)(x + 7) = 0$

14. $(5y - 1)(2y + 4) = 0$

15. $(4m + 16)(3m - 18) = 0$

16. $(2n - 7)(5n + 20) = 0$

17. $(2x - 5)(9x - 15) = 0$

18. $(7t + 21)(t + 9) = 0$

19. $(5x + 7)(7x - 15) = 0$

20. $(2c + 8)(5c + 10) = 0$

21. $(2w - 6.4)(4w - 8.4) = 0$

22. $(3x + 9.3)(4x - 12.8) = 0$

23. $\left(2x - \frac{1}{2}\right)\left(2x + \frac{1}{2}\right)^2 = 0$

24. $\left(5n + \frac{1}{3}\right)\left(3n - \frac{1}{2}\right) = 0$

Match the function with its graph.

A. $y = (x + 4)(x - 1)$

B. $y = (x - 4)(x + 1)$

C. $y = (x - 4)(x - 1)$

25.

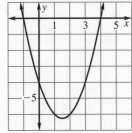

26.

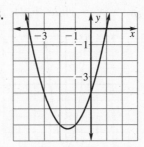

27.

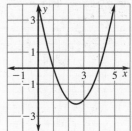

**Find the *x*-intercepts and the vertex of the graph of the function.
Then sketch the graph of the function.**

28. $y = (x - 4)(x + 2)$

29. $y = (x - 6)(x + 4)$

30. $y = (x + 1)(x - 5)$

31. $y = (-x - 6)(x - 6)$

32. $y = (x - 3)(-x + 7)$

33. $y = (x - 5)(x - 3)$

34. A diver jumps from a diving board that is 32 feet above the water.
The height of the diver is given by

$$h = -16(t - 2)(t + 1)$$

where the height *h* is measured in feet, and the time *t* is measured in
seconds. When will the diver hit the water? Can you see a quick way to
find the answer? Explain.

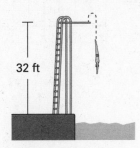

32 ft

NAME _____ DATE _____

Practice C

For use with pages 597–602

Use the zero-product property to solve the equation.

1. $(x + 2)(x + 7) = 0$

2. $(t + 9)(t - 9) = 0$

3. $(x + 4)(x - 8) = 0$

4. $(c + 5)(c + 1) = 0$

5. $(n - 9)(n - 3) = 0$

6. $(x + 7.2)(x - 7.2) = 0$

7. $\left(d + \frac{3}{5}\right)\left(d - \frac{3}{5}\right) = 0$

8. $\left(x + \frac{2}{3}\right)\left(x - \frac{2}{3}\right) = 0$

9. $(x + 6.1)(x - 5.3) = 0$

10. $5(m + 6)^2 = 0$

11. $(w - 5.2)\left(w - \frac{5}{2}\right) = 0$

12. $(y - 4)(y + 4)^2 = 0$

Solve the equation.

13. $(2x + 6)(x + 6) = 0$

14. $(5y - 3)(2y + 5) = 0$

15. $(4m + 12)(3m - 12) = 0$

16. $(7n - 2)(5n + 15) = 0$

17. $(2x - 7.8)(9x - 12.6) = 0$

18. $(21t + 7)(9t + 1) = 0$

19. $(4x + 3.2)(9x - 24.3) = 0$

20. $(2c + 4)(5c + 5)(c - 4) = 0$

21. $(2w - 6)(4w - 8)(w + 1) = 0$

22. $(3x + 6.9)^2(4x - 16.4) = 0$

23. $\left(3x - \frac{1}{2}\right)\left(3x + \frac{1}{2}\right)^2 = 0$

24. $\left(6n + \frac{1}{2}\right)\left(6n - \frac{1}{3}\right) = 0$

Match the function with its graph.

A. $y = (2x + 4)(x - 2)$

B. $y = (x - 4)(2x + 2)$

C. $y = (2x - 2)(x + 4)$

25.

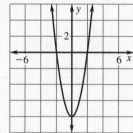

26.

27.

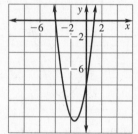

Find the *x*-intercepts and the vertex of the graph of the function. Then sketch the graph of the function.

28. $y = (x - 6)(x + 4)$

29. $y = (x - 1)(x + 5)$

30. $y = (x + 3)(x - 5)$

31. $y = (-x - 3)(x - 4)$

32. $y = (x - 2)(-x + 5)$

33. $y = (x - 5)(x - 2)$

40. The cross section of a wooden storage structure can be modeled by the polynomial function

$$y = \frac{-15}{400}(2x - 35)(2x + 35)$$

where x and y are measured in feet, and the center of the structure is where $x = 0$. Explain how to use the algebraic model to find the width of the structure. What is the structure's width? Use the model to find the coordinates of the center of the structure.

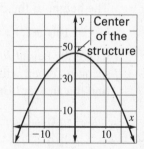

NAME _____ DATE _____

Reteaching with Practice

For use with pages 597–602

GOAL **Solve a polynomial equation in factored form and relate factors and**
x-intercepts

> ### VOCABULARY
>
> A polynomial is in **factored form** if it is written as the product of two
> or more linear factors. According to the **zero-product property,** the
> product of two factors is zero, only when at least one of the factors is
> zero.

EXAMPLE 1 *Using the Zero-Product Property*

Solve the equation $(x - 1)(x + 7) = 0$.

SOLUTION

Use the zero-product property: either $x - 1 = 0$ or $x + 7 = 0$.

$(x - 1)(x + 7) = 0$	Write original equation.
$x - 1 = 0$	Set first factor equal to 0.
$x = 1$	Solve for x.
$x + 7 = 0$	Set second factor equal to 0.
$x = -7$	Solve for x.

The solutions are 1 and -7.

Exercises for Example 1

Solve the equation.

1. $(z - 6)(z + 6) = 0$ **2.** $(y - 5)(y - 1) = 0$ **3.** $(x + 4)(x + 3) = 0$

EXAMPLE 2 *Using the Zero-Product Property*

Solve the equation $(x - 4)^2 = 0$.

SOLUTION

This equation has a repeated factor. To solve the equation, you need to
set only $x - 4$ equal to zero.

$(x - 4)^2 = 0$	Write original equation.
$x - 4 = 0$	Set repeated factor equal to 0.
$x = 4$	Solve for x.

The solution is 4.

Lesson 10.4

NAME _____ DATE _____

Reteaching with Practice

For use with pages 597–602

Exercises for Example 2

Solve the equation.

4. $(t - 5)^2 = 0$ **5.** $(y + 3)^2 = 0$ **6.** $(2x + 4)^2 = 0$

EXAMPLE 3   *Relating x-Intercepts and Factors*

Name the *x*-intercepts and the vertex of the graph
of the function $y = (x + 4)(x - 2)$. Then sketch
the graph of the function.

SOLUTION

First solve $(x + 4)(x - 2) = 0$ to find the
x-intercepts: -4 and 2.

Then find the coordinates of the vertex.

• The *x*-coordinate of the vertex is the average
  of the *x*-intercepts.

$$x = \frac{-4 + 2}{2} = -1$$

• Substitute to find the *y*-coordinate.

$$y = (-1 + 4)(-1 - 2) = -9$$

• The coordinates of the vertex are $(-1, -9)$

Exercises for Example 3

**Name the *x*-intercepts and the vertex of the graph of the
function.**

7. $y = (x + 3)(x + 1)$ **8.** $y = (x - 2)(x - 4)$ **9.** $y = (x - 1)(x + 5)$

NAME _____ DATE _____

Quick Catch-Up for Absent Students

For use with pages 597–602

The items checked below were covered in class on (date missed) _____

Lesson 10.4: Solving Polynomial Equations in Factored Form

____ **Goal 1:** Solve a polynomial equation in factored form. (pp. 597–598)

Material Covered:

____ Activity: Investigating Factored Equations

____ Example 1: Using the Zero-Product Property

____ Example 2: Solving a Repeated-Factor Equation

____ Student Help: Study Tip

____ Example 3: Solving a Factored Cubic Equation

Vocabulary:

factored form of a polynomial, p. 597   zero-product property, p. 597

____ **Goal 2:** Relate factors and *x*-intercepts. (p. 599)

Material Covered:

____ Student Help: Study Tip

____ Example 4: Relating *x*-Intercepts and Factors

____ Example 5: Using a Quadratic Model

____ Other (specify) _____

_____

Homework and Additional Learning Support

____ Textbook (specify) _pp. 600–602_____

_____

____ *Reteaching with Practice* worksheet (specify exercises)_____

____ *Personal Student Tutor* for Lesson 10.4

Real-Life Application:
When Will I Ever Use This?

For use with pages 596–602

Track and Field

Track and field is a sport in which athletes compete in running, walking, jumping, and throwing events. Track events consist of running and walking races of various distances. Field events are contests in jumping and throwing.

Javelin is a light, slender spear that is thrown for distance. In ancient times, warriors used the javelin as a weapon of war. Ancient hunters also used the javelin. The javelin used in track and field meets is made of metal with a metal tip. The length of the men's javelin ranges from 2.6 to 2.7 meters. The length of the women's javelin ranges from 2.2 to 2.3 meters.

In Exercises 1-3, use the following information.

A female Olympic athlete throws a javelin that follows a path modeled by $y = -\frac{1}{605}(x - 110)(x + 110)$, with x and y measured in feet.

1. Find the x-intercepts and sketch the graph of the javelin throw. Let the y-axis represent the maximum height of the javelin.

2. Estimate the horizontal distance the javelin travels.

3. Estimate the height the javelin reaches.

The shot-put is a throwing event in which athletes put (push) a metal ball called a shot. They hold the heavy shot against their neck and release it with a powerful push. The men's shot weighs at least 16 pounds. The women's shot weighs at least 8 pounds 13 ounces.

In Exercises 4-6, use the following information.

A male Olympic athlete hurtles a shot that follows a path modeled by $y = -\frac{1}{56}(x - 28)(x + 28)$, with x and y measured in feet.

4. Find the x-intercepts and sketch the graph of the shot-put throw. Let the y-axis represent the maximum height of the shot.

5. Estimate the horizontal distance and the height of the shot.

6. In Exercises 2, 3, and 5, you estimated the horizontal distances and heights. Why are these measurements estimated?

NAME _____ DATE _____

Challenge: Skills and Applications

For use with pages 597–602

In Exercises 1–6, solutions are given. Find a polynomial equation that has only these solutions.

1. $0, 3, -2$

2. 0 (repeated factor), -4

3. $-\frac{1}{2}, 5, 7$

4. $\frac{2}{3}, 4, -\frac{5}{6}$

5. $-k, \frac{k}{5}, 4k$

6. $2k, \frac{4k}{7}, -\frac{5}{9k}$

In Exercises 7–8, use the following information.

A poster is 12 inches wide and 20 inches high. The border of the poster is twice as wide on the top and bottom as it is on the sides. The printed part of the poster has an area of 126 square inches.

7. Write an algebraic model for this situation.

8. Find the width of the side borders and the width of the top and bottom borders.

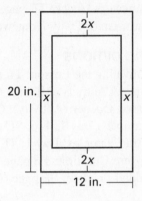

In Exercises 9–12, use the following information.

An on-line computer service company charges a monthly fee of $12 and has 6000 customers. A market survey has shown that for every $1 increase in the monthly fee the company would lose 200 customers.

9. Suppose the monthly fee is increased by n dollars above its present figure. Write expressions for the new monthly fee, the new number of customers the company would have, and the total monthly revenue that the company would take in (number of customers × monthly fee).

10. Graph the revenue function from Exercise 9.

11. For what monthly fee would the company have no customers at all?

12. What monthly fee should the company charge in order to maximize its monthly revenue?

TEACHER'S NAME _____ CLASS _____ ROOM _____ DATE _____

Lesson Plan

2-day lesson (See *Pacing the Chapter,* TE pages 572C–572D)   For use with pages 603–609

GOALS 1. Factor a quadratic expression of the form $x^2 + bx + c$.
2. Solve quadratic equations by factoring.

State/Local Objectives _____

_____

✓ **Check the items you wish to use for this lesson.**

STARTING OPTIONS
____ Homework Check: TE page 600; Answer Transparencies
____ Warm-Up or Daily Homework Quiz: TE pages 604 and 602, CRB page 66, or Transparencies

TEACHING OPTIONS
____ Motivating the Lesson: TE page 605
____ Concept Activity: SE page 603; CRB page 67 (Activity Support Master)
____ Lesson Opener (Activity): CRB page 68 or Transparencies
____ Examples: Day 1: 1–6, SE pages 604–606; Day 2: 7, SE page 606
____ Extra Examples: Day 1: TE pages 605–606 or Transp.; Day 2: TE page 606 or Transp.; Internet
____ Closure Question: TE page 606
____ Guided Practice: SE page 607; Day 1: Exs. 1–11; Day 2: none

APPLY/HOMEWORK
Homework Assignment
____ Basic   Day 1: 12–46 even, 52, 53, 56–58; Day 2: 13–47 odd, 54, 55, 59, 60, 65–67, 75, 80, 85, 90, 95, 96
____ Average   Day 1: 12–46 even, 52, 53, 56–58; Day 2: 13–47 odd, 54, 55, 59–62, 65–67, 75, 80, 85, 90, 95, 96
____ Advanced   Day 1: 12–46 even, 52, 53, 56–58; Day 2: 13–47 odd, 54, 55, 59–62, 65–71, 75, 80, 85, 90, 95, 96

Reteaching the Lesson
____ Practice Masters: CRB pages 69–71 (Level A, Level B, Level C)
____ Reteaching with Practice: CRB pages 72–73 or Practice Workbook with Examples
____ Personal Student Tutor

Extending the Lesson
____ Applications (Interdisciplinary): CRB page 75
____ Challenge: SE page 609; CRB page 76 or Internet

ASSESSMENT OPTIONS
____ Checkpoint Exercises: Day 1: TE pages 605–606 or Transp.; Day 2: TE page 606 or Transp.
____ Daily Homework Quiz (10.5): TE page 609, CRB page 79, or Transparencies
____ Standardized Test Practice: SE page 609; TE page 609; STP Workbook; Transparencies

Notes _____

_____

_____

TEACHER'S NAME _____ CLASS _____ ROOM _____ DATE _____

Lesson Plan for Block Scheduling

1-day lesson (See *Pacing the Chapter*, TE pages 572C–572D)　　　　For use with pages 603–609

GOALS　1. **Factor a quadratic expression of the form** $x^2 + bx + c$.
　　　　　2. **Solve quadratic equations by factoring.**

State/Local Objectives _____

_____

_____

✓ **Check the items you wish to use for this lesson.**

STARTING OPTIONS
____ Homework Check: TE page 600; Answer Transparencies
____ Warm-Up or Daily Homework Quiz: TE pages 604 and
　　　　602, CRB page 66, or Transparencies

TEACHING OPTIONS
____ Motivating the Lesson: TE page 605
____ Concept Activity: SE page 603; CRB page 67 (Activity Support Master)
____ Lesson Opener (Activity): CRB page 68 or Transparencies
____ Examples 1–7: SE pages 604–606
____ Extra Examples: TE pages 605–606 or Transparencies; Internet
____ Closure Question: TE page 606
____ Guided Practice Exercises: SE page 607

APPLY/HOMEWORK
Homework Assignment
____ Block Schedule: 12–47, 52–62, 65–67, 75, 80, 85, 90, 95, 96

Reteaching the Lesson
____ Practice Masters: CRB pages 69–71 (Level A, Level B, Level C)
____ Reteaching with Practice: CRB pages 72–73 or Practice Workbook with Examples
____ Personal Student Tutor

Extending the Lesson
____ Applications (Interdisciplinary): CRB page 75
____ Challenge: SE page 609; CRB page 76 or Internet

ASSESSMENT OPTIONS
____ Checkpoint Exercises: TE pages 605–606 or Transparencies
____ Daily Homework Quiz (10.5): TE page 609, CRB page 79, or Transparencies
____ Standardized Test Practice: SE page 609; TE page 609; STP Workbook; Transparencies

Notes _____

_____

_____

CHAPTER PACING GUIDE	
Day	**Lesson**
1	Assess Ch. 9; 10.1 (all)
2	10.2 (all); 10.3 (begin)
3	10.3 (end); 10.4 (all)
4	**10.5 (all)**
5	10.6 (all)
6	10.7 (all)
7	10.8 (all)
8	Review/Assess Ch. 10

WARM-UP EXERCISES

For use before Lesson 10.5, pages 603–609

Find the product.

1. $(x + 2)(x + 6)$

2. $(x - 9)(x + 9)$

3. $(x - 3)^2$

4. $(2x - 5)(x + 5)$

5. $\left(x + \dfrac{1}{4}\right)\left(x - \dfrac{1}{4}\right)$

DAILY HOMEWORK QUIZ

For use after Lesson 10.4, pages 597–602

Solve the equation.

1. $(y - 7)^2 = 0$

2. $(x + 3)(x - 9) = 0$

3. $2(2t - 2)(3t + 5) = 0$

4. Find the x-intercepts and the vertex of the graph of
$y = (x + 1)(x - 1)$. Then sketch the graph of the function.

NAME _____ DATE _____

Activity Support Master

For use with page 603

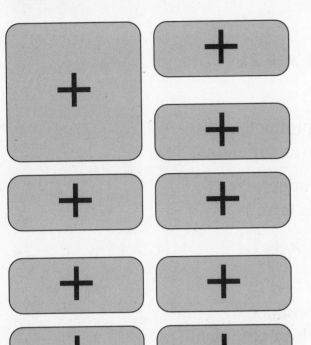

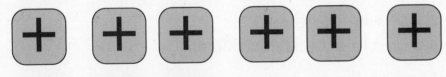

Activity Lesson Opener

For use with pages 604–609

SET UP: Work with a partner.

Consider the binomials $x + 3$ and $x + 2$.

1. Find the product of the binomials.

2. Look at the coefficient of x in the product.
   How did you use 3 and 2 to get this coefficient?

3. Look at the constant term in the product.
   How did you use 3 and 2 to get this term?

Consider the binomials $x - 4$ and $x + 1$.

4. Find the product of the binomials.

5. Look at the coefficient of x in the product.
   How did you use -4 and 1 to get this coefficient?

6. Look at the constant term in the product.
   How did you use -4 and 1 to get this term?

Consider the binomials $x + 6$ and $x - 3$.

7. Find the product of the binomials.

8. Look at the coefficient of x in the product.
   How did you use 6 and -3 to get this coefficient?

9. Look at the constant term in the product.
   How did you use 6 and -3 to get this term?

Consider the binomials $x - 3$ and $x - 5$.

10. Find the product of the binomials.

11. Look at the coefficient of x in the product. How
    did you use -3 and -5 to get this coefficient?

12. Look at the constant term in the product.
    How did you use -3 and -5 to get this term?

13. Make a statement about the relationships you have
    discovered.

Algebra 1
Chapter 10 Resource Book

LESSON 10.5

Practice A

For use with pages 604–609

Use the model to write the factors of the trinomial.

1.
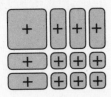

2.

Match the trinomial with a correct factorization.

3. $x^2 - 5x + 6$ **A.** $(x + 3)(x + 2)$

4. $x^2 + 5x + 6$ **B.** $(x - 3)(x + 2)$

5. $x^2 - x - 6$ **C.** $(x + 3)(x - 2)$

6. $x^2 + x - 6$ **D.** $(x - 3)(x - 2)$

Factor the trinomial.

7. $x^2 + 6x + 8$ 8. $x^2 + 3x - 4$ 9. $x^2 + 3x + 2$

10. $x^2 - 2x - 8$ 11. $x^2 + 7x + 12$ 12. $x^2 - 6x + 5$

13. $x^2 + x - 20$ 14. $x^2 + 8x + 16$ 15. $x^2 - 10x + 24$

Solve the equation by factoring.

16. $x^2 + 3x - 4 = 0$ 17. $x^2 - 5x + 6 = 0$ 18. $x^2 + 3x - 18 = 0$

19. $x^2 - 16x - 36 = 0$ 20. $x^2 + 8x + 7 = 0$ 21. $x^2 + 3x - 10 = 0$

22. $x^2 + 5x = 14$ 23. $x^2 - 7x = 8$ 24. $x^2 - 9x + 20 = 0$

25. $x^2 - 2x - 48 = 0$ 26. $x^2 + 12x = -27$ 27. $x^2 + 3x = 28$

Find the dimensions of the geometric shape.

28.
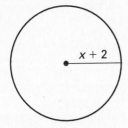

Area $= 144\pi$ cm^2

29.
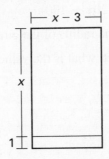

Area $= 60$ in.^{2}

30.
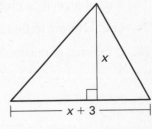

Area $= 27$ in.^{2}

NAME _____ DATE _____

Practice B

For use with pages 604–609

Match the trinomial with a correct factorization.

1. $x^2 - 5x + 6$
2. $x^2 + 5x + 6$
3. $x^2 - x - 6$
4. $x^2 + x - 6$
5. $x^2 - 4x + 4$
6. $x^2 - 6x + 9$

A. $(x - 2)(x - 2)$
B. $(x - 3)(x + 2)$
C. $(x - 3)(x - 3)$
D. $(x - 3)(x - 2)$
E. $(x + 3)(x + 2)$
F. $(x + 3)(x - 2)$

Factor the trinomial.

7. $x^2 - 5x - 14$
8. $x^2 - 8x + 15$
9. $x^2 + 8x + 15$
10. $x^2 - 5x + 4$
11. $x^2 - x - 42$
12. $x^2 + 6x - 16$
13. $x^2 - 16x + 64$
14. $x^2 + 13x + 36$
15. $x^2 - 15x + 36$

Solve the equation by factoring.

16. $x^2 + 3x - 40 = 0$
17. $x^2 - 16x + 63 = 0$
18. $x^2 - 11x + 28 = 0$
19. $x^2 - 6x - 7 = 0$
20. $x^2 - 6x + 9 = 0$
21. $x^2 + 8x + 15 = 0$
22. $x^2 + x = 6$
23. $x^2 + 11x = 12$
24. $x^2 - 3x = 28$
25. $x^2 - 7 = -6x$
26. $x^2 - 8 = -7x$
27. $x^2 - 4x - 8 = 4$

Tell whether the quadratic expression can be factored with integer coefficients. If it can, find the factors.

28. $x^2 + 17x + 60$
29. $x^2 - 15x + 48$
30. $x^2 - 5x - 36$
31. $x^2 + 13x + 30$
32. $x^2 + 11x + 30$
33. $x^2 + 8x - 40$

Area of a Circle   **In Exercises 34 and 35, use the following information.**

The area of a circle is given by $A = \pi(x^2 - 20x + 100)$.

34. Use factoring to find an expression for the radius of the circle.

35. If the area of the circle is 16π square feet, what is the value of x?

NAME _____   DATE _____

Practice C
For use with pages 604–609

Factor the trinomial.

1. $x^2 + 5x + 6$

2. $x^2 + 6x + 8$

3. $x^2 - 4x + 3$

4. $x^2 - 11x + 30$

5. $x^2 - 2x - 8$

6. $x^2 - x - 12$

7. $x^2 + 3x - 28$

8. $x^2 + 5x - 14$

9. $x^2 + 8x + 15$

10. $x^2 - 20x + 100$

11. $x^2 + 17x + 72$

12. $x^2 - 12x - 64$

Solve the equation by factoring.

13. $x^2 - 13x + 36 = 0$

14. $x^2 - 3x - 70 = 0$

15. $x^2 + 4x - 45 = 0$

16. $x^2 + 11x + 28 = 0$

17. $x^2 - 15x + 44 = 0$

18. $x^2 + 3x = 18$

19. $x^2 - 2x = 63$

20. $x^2 - 14 = 5x$

21. $x^2 + 10 = 11x$

22. $x^2 - x = 12$

23. $x^2 - 4x = -3$

24. $x^2 - 14 = -5x$

25. $x^2 - x = 3x + 12$

26. $x^2 + 6x + 10 = 2$

27. $x^2 + 2x - 40 = 40$

Use the discriminant to tell whether the quadratic expression can be factored with integer coefficients. If it can, find the factors.

28. $x^2 - 12x + 32$

29. $x^2 - 13x - 48$

30. $x^2 - x - 90$

31. $x^2 - 5x - 84$

32. $x^2 - 17x + 66$

33. $x^2 + 10x - 44$

Write a quadratic equation that has the given solutions.

34. 12 and 5

35. -18 and 20

36. 25 and 0

Area of a Rectangle   **In Exercises 37–39, use the following information.**

The area of a rectangle is given by $A = x^2 + 18x + 72$.

37. Use factoring to find an expression for the dimensions of the rectangle.

38. If the area of the rectangle is 7 square feet, what are the possible values of x?

39. What are the dimensions of the rectangle?

NAME _____ DATE _____

Reteaching with Practice

For use with pages 604–609

GOAL Factor a quadratic expression of the form $x^2 + bx + c$ and solve quadratic equations by factoring

---

VOCABULARY

To **factor** a quadratic expression means to write it as the product of two linear expressions. To factor $x^2 + bx + c$, you need to find numbers p and q such that

$$p + q = b \quad \text{and} \quad pq = c.$$

$x^2 + bx + c = (x + p)(x + q)$ when $p + q = b$ and $pq = c$

EXAMPLE 1 *Factoring when b and c are Positive*

Factor $x^2 + 6x + 8$.

SOLUTION

For this trinomial, $b = 6$ and $c = 8$. You need to find two numbers whose sum is 6 and whose product is 8.

$$x^2 + 6x + 8 = (x + p)(x + q) \qquad \text{Find } p \text{ and } q \text{ when } p + q = 6 \text{ and } pq = 8.$$
$$= (x + 4)(x + 2) \qquad p = 4 \text{ and } q = 2$$

Exercises for Example 1

Factor the trinomial.

1. $x^2 + 5x + 6$　　　　　　　**2.** $x^2 + 6x + 5$　　　　　　　**3.** $x^2 + 3x + 2$

EXAMPLE 2 *Factoring when b is Negative and c is Positive*

Factor $x^2 - 5x + 4$.

SOLUTION

Because b is negative and c is positive, both p and q must be negative numbers. Find two numbers whose sum is -5 and whose product is 4.

$$x^2 - 5x + 4 = (x + p)(x + q) \qquad \text{Find } p \text{ and } q \text{ when } p + q = -5 \text{ and } pq = 4.$$
$$= (x - 4)(x - 1) \qquad p = -4 \text{ and } q = -1$$

Exercises for Example 2

Factor the trinomial.

4. $x^2 - 3x + 2$　　　　　　**5.** $x^2 - 7x + 12$　　　　　　**6.** $x^2 - 5x + 6$

Reteaching with Practice

For use with pages 604–609

EXAMPLE 3 *Factoring when b and c are Negative*

Factor $x^2 - 3x - 10$.

SOLUTION

For this trinomial, $b = -3$ and $c = -10$. Because c is negative, you know that p and q cannot both have negative values.

$$x^2 - 3x - 10 = (x + p)(x + q) \quad \text{Find } p \text{ and } q \text{ when } p + q = -3 \text{ and } pq = -10.$$
$$= (x + 2)(x - 5) \quad p = 2 \text{ and } q = -5$$

Exercises for Example 3

Factor the trinomial.

7. $x^2 - x - 2$ **8.** $x^2 - 4x - 12$ **9.** $x^2 - 2x - 8$

EXAMPLE 4 *Solving a Quadratic Equation*

Solve $x^2 + 4x = 12$.

SOLUTION

$x^2 + 4x = 12$	Write equation.
$x^2 + 4x - 12 = 0$	Write in standard form.
$(x + 6)(x - 2) = 0$	Factor left side. Because c is negative, p and q cannot both have negative values: $p = 6$ and $q = -2$
$(x + 6) = 0 \text{ or } (x - 2) = 0$	Use zero-product property.
$x + 6 = 0$	Set first factor equal to 0.
$x = -6$	Solve for x.
$x - 2 = 0$	Set second factor equal to 0.
$x = 2$	Solve for x.

The solutions are -6 and 2.

Exercises for Example 4

Solve the equation.

10. $x^2 + 8x + 15 = 0$ **11.** $x^2 - 8x + 12 = 0$ **12.** $x^2 + 3x - 4 = 0$

NAME _____ DATE _____

Quick Catch-Up for Absent Students

For use with pages 603–609

The items checked below were covered in class on (date missed) _____

Activity 10.5: Modeling the Factorization of $x^2 + bx + c$ (p. 603)

_____ **Goal:** Model the factorization of a trinomial of the form $x^2 + bx + c$ using algebra tiles.

Lesson 10.5: Factoring $x^2 + bx + c$

_____ **Goal 1:** Factor a quadratic expression of the form $x^2 + bx + c$. (pp. 604–605)

Material Covered:

_____ Example 1: Factoring when b and c are Positive

_____ Example 2: Factoring when b is Negative and c is Positive

_____ Example 3: Factoring when b and c are Negative

_____ Example 4: Factoring when b is Positive and c is Negative

_____ Student Help: Look Back

_____ Example 5: Using the Discriminant

Vocabulary:

factor a quadratic expression, p. 604

_____ **Goal 2:** Solve quadratic equations by factoring. (p. 606)

Material Covered:

_____ Example 6: Solving a Quadratic Equation

_____ Example 7: Writing a Quadratic Model

_____ Other (specify) _____
_____

Homework and Additional Learning Support

_____ Textbook (specify) _pp. 607–609_____
_____

_____ Internet: Extra Examples at www.mcdougallittell.com

_____ *Reteaching with Practice* worksheet (specify exercises)_____

_____ *Personal Student Tutor* for Lesson 10.5

NAME _____ DATE _____

Interdisciplinary Application

For use with pages 604–609

Marching Band

MUSIC Many junior highs, high schools, and colleges offer marching band as an extracurricular activity for students. These bands perform and compete at parades, athletic competitions and other outdoor events.

Your marching band is asked to perform in a local parade. The band is allotted 3200 square feet in the parade and must stay at least 5 feet from the curb. The width of the street is x feet and the length of space allotted for the band is $x + 30$ feet.

Your director decides that rows will be 4 feet apart and each row will contain 8 band members.

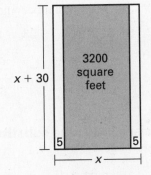

1. Use the diagram to write a polynomial expression representing the area the band is allotted.

2. Simplify the expression in Exercise 1.

3. Find x by setting the simplified expression in Exercise 2 equal to the square footage allotted for the band.

4. Find the number of rows your band will take up. (*Hint:* row 1 starts at 0 feet, row 2 is at 4 feet, and so on.)

5. Can all 200 members be in the parade? If not, how many more square feet are needed?

Challenge: Skills and Applications

For use with pages 604–609

In Exercises 1–6, use substitution to factor the expression.

Example: $x + 3x^{1/2} - 28$

Solution: Since $x = (x^{1/2})^2$, you can rewrite the expression as
$(x^{1/2})^2 + 3x^{1/2} - 28$. Then you can do the substitution $y = x^{1/2}$.
The polynomial becomes $y^2 + 3y - 28 = (y - 4)(y + 7)$.

So, $x + 3x^{1/2} - 28 = (x^{1/2} - 4)(x^{1/2} + 7)$.

1. $(x^{1/2})^2 - 7x^{1/2} + 10$

2. $x^6 - 4x^3 - 12$

3. $x - 4\sqrt{x} - 45$

4. $x + 11\sqrt{x} + 10$

5. $\dfrac{1}{x^2} - \dfrac{15}{x} + 56$

6. $\dfrac{1}{x^2} + \dfrac{3}{x} - 108$

In Exercises 7–10, use substitution to factor and then solve the equation.

7. $x^4 - 13x^2 + 36 = 0$

8. $x^4 - 11x^2 + 24 = 0$

9. $\dfrac{1}{x^2} + \dfrac{1}{6x} - \dfrac{1}{6} = 0$

10. $\dfrac{1}{x^2} + \dfrac{13}{(3x)} + \dfrac{10}{3} = 0$

11. How many solutions does $x^4 - 13x^2 - 48$ have? What are they?

In Exercises 12–14, use the following information.

A square garden has a sidewalk around its outer edge. The sidewalk is a little wider on one set of opposite sides than on the other set. The area inside the sidewalk is given by the trinomial $x^2 - 13.6x + 46.2$, where x is the length and width of the entire garden, in feet.

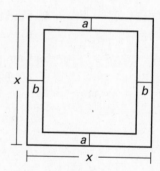

12. Factor $x^2 - 13.6x + 46.2$ (*Hint:* Use the quadratic formula or a graph.)

13. What are the widths of the sidewalks?

14. If the area inside the sidewalks is 2830.2 square feet, what are the dimensions of the garden?

TEACHER'S NAME _____ CLASS _____ ROOM _____ DATE _____

Lesson Plan

2-day lesson (See *Pacing the Chapter,* TE pages 572C–572D)   **For use with pages 610–618**

GOALS   1. **Factor a quadratic expression of the form** $ax^2 + bx + c$.
2. **Solve quadratic equations by factoring.**

State/Local Objectives _____

_____

✓ **Check the items you wish to use for this lesson.**

STARTING OPTIONS

____ Homework Check: TE page 607; Answer Transparencies
____ Warm-Up or Daily Homework Quiz: TE pages 611 and 609, CRB page 79, or Transparencies

TEACHING OPTIONS

____ Motivating the Lesson: TE page 612
____ Concept Activity: SE page 610; CRB page 80 (Activity Support Master)
____ Lesson Opener (Graphing Calculator): CRB page 81 or Transparencies
____ Graphing Calculator Activity with Keystrokes: CRB page 82
____ Examples: Day 1: 1–5, SE pages 611–613; Day 2: 6, SE page 613
____ Extra Examples: Day 1: TE pages 612–613 or Transp.; Day 2: TE page 613 or Transp.
____ Technology Activity: SE page 618
____ Closure Question: TE page 613
____ Guided Practice: SE page 614; Day 1: Exs. 1–3, 5–14; Day 2: Ex. 4

APPLY/HOMEWORK

Homework Assignment

____ Basic   Day 1: 16–62 even; Day 2: 15–63 odd, 64, 65, 73–76, 80, 83, 85, 88, 90, Quiz 2: 1–27
____ Average   Day 1: 16–62 even; Day 2: 15–63 odd, 64, 65, 73–76, 80, 83, 85, 88, 90, Quiz 2: 1–27
____ Advanced   Day 1: 16–62 even; Day 2: 15–63 odd, 64, 65, 73–76, 80, 83, 85, 88, 90, Quiz 2: 1–27

Reteaching the Lesson

____ Practice Masters: CRB pages 83–85 (Level A, Level B, Level C)
____ Reteaching with Practice: CRB pages 86–87 or Practice Workbook with Examples
____ Personal Student Tutor

Extending the Lesson

____ Applications (Interdisciplinary): CRB page 89
____ Challenge: SE page 616; CRB page 90 or Internet

ASSESSMENT OPTIONS

____ Checkpoint Exercises: Day 1: TE pages 612–613 or Transp.; Day 2: TE page 613 or Transp.
____ Daily Homework Quiz (10.6): TE page 616, CRB page 94, or Transparencies
____ Standardized Test Practice: SE page 616; TE page 616; STP Workbook; Transparencies
____ Quiz (10.4–10.6): SE page 617; CRB page 91

Notes _____

_____

_____

Lesson 10.6

TEACHER'S NAME _____ CLASS _____ ROOM _____ DATE _____

Lesson Plan for Block Scheduling

1-day lesson (See *Pacing the Chapter,* TE pages 572C–572D)   For use with pages 610–618

 GOALS   1. **Factor a quadratic expression of the form** $ax^2 + bx + c$.
2. **Solve quadratic equations by factoring.**

State/Local Objectives _____

_____

_____

✓ Check the items you wish to use for this lesson.

STARTING OPTIONS

_____ Homework Check: TE page 607; Answer Transparencies

_____ Warm-Up or Daily Homework Quiz: TE pages 611 and
        609, CRB page 79, or Transparencies

TEACHING OPTIONS

_____ Motivating the Lesson: TE page 612

_____ Concept Activity: SE page 610; CRB page 80 (Activity Support Master)

_____ Lesson Opener (Graphing Calculator): CRB page 81 or Transparencies

_____ Graphing Calculator Activity with Keystrokes: CRB page 82

_____ Examples 1–6: SE pages 611–613

_____ Extra Examples: TE pages 612–613 or Transparencies

_____ Technology Activity: SE page 618

_____ Closure Question: TE page 613

_____ Guided Practice Exercises: SE page 614

APPLY/HOMEWORK

Homework Assignment

_____ Block Schedule: 15–65, 73–76, 80, 83, 85, 88, 90, Quiz 2: 1–27

Reteaching the Lesson

_____ Practice Masters: CRB pages 83–85 (Level A, Level B, Level C)

_____ Reteaching with Practice: CRB pages 86–87 or Practice Workbook with Examples

_____ Personal Student Tutor

Extending the Lesson

_____ Applications (Interdisciplinary): CRB page 89

_____ Challenge: SE page 616; CRB page 90 or Internet

ASSESSMENT OPTIONS

_____ Checkpoint Exercises: TE pages 612–613 or Transparencies

_____ Daily Homework Quiz (10.6): TE page 616, CRB page 94, or Transparencies

_____ Standardized Test Practice: SE page 616; TE page 616; STP Workbook; Transparencies

_____ Quiz (10.4–10.6): SE page 617; CRB page 91

Notes _____

_____

_____

CHAPTER PACING GUIDE	
Day	**Lesson**
1	Assess Ch. 9; 10.1 (all)
2	10.2 (all); 10.3 (begin)
3	10.3 (end); 10.4 (all)
4	10.5 (all)
5	**10.6 (all)**
6	10.7 (all)
7	10.8 (all)
8	Review/Assess Ch. 10

Lesson 10.6

NAME _____ DATE _____

WARM-UP EXERCISES

For use before Lesson 10.6, pages 610–618

Factor the trinomial.

1. $x^2 - 10x + 24$

2. $x^2 + 10x + 9$

3. $x^2 + 6x - 40$

4. $x^2 - 6x - 40$

5. $x^2 - 11x + 24$

DAILY HOMEWORK QUIZ

For use after Lesson 10.5, pages 603–609

Factor the trinomial.

1. $y^2 + 5y - 14$

2. $x^2 - 3x - 88$

3. $t^2 + 23t + 90$

4. Solve the equation $x^2 - 32 = 4x$.

5. Can $y^2 + 2y - 120$ be factored with integer coefficients? If so, what are the factors?

Lesson 10.6

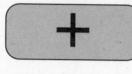

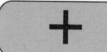

Algebra 1
Chapter 10 Resource Book

NAME _____ DATE _____

Graphing Calculator Lesson Opener

For use with pages 611–617

1. Use your calculator to graph $y = 2x^2 + 7x + 3$.

2. Do not clear the screen. Graph $y = (2x + 3)(x + 1)$ on the same screen as the function in Question 1. Describe the results.

3. Multiply the first terms of the two binomials in Question 2. Compare the result to the first term of the trinomial in Question 1.

4. Multiply the last terms of the two binomials in Question 2. Compare the result to the third term of the trinomial in Question 1.

5. Use FOIL to find the middle term of the product of the two binomials in Question 2. Compare the result to the middle term of the trinomial in Question 1.

6. Clear the function you graphed in Question 2, but not the function you graphed in Question 1. Graph $y = (2x + 1)(x + 3)$. Describe the results.

7. Multiply the first terms of the two binomials in Question 6. Compare the result to the first term of the trinomial in Question 1.

8. Multiply the last terms of the two binomials in Question 6. Compare the result to the third term of the trinomial in Question 1.

9. Use FOIL to find the middle term of the product of the two binomials in Question 6. Compare the result to the middle term of the trinomial in Question 1.

10. What conclusion can you make about the relationships between the trinomial in Question 1 and the binomial products in each of Questions 2 and 6?

Graphing Calculator Activity

For use with pages 611–617

Goal: To program a graphing calculator to find the value of a discriminant.

You can program a graphing calculator or computer to find the value of the discriminant of $ax^2 + bx + c$. If the discriminant is not a perfect square, then $ax^2 + bx + c = 0$ cannot be solved by factoring with integer coefficients.

Activity

1. Enter the program below that is written for your calculator.

TI-82

PROGRAM: DISC
: Prompt A, B, C
: $B^2 - 4AC \rightarrow D$
: If D < 0
: Then
: Disp "NO SOLUTION"
: Else
: Disp "$B^2 - 4AC = $"
: Disp D
: Pause
: Disp "$\sqrt{(B^2 - 4AC)} = $"
: Disp $\sqrt{D}$
: End

TI-83

PROGRAM: DISC
: Prompt A, B, C
: $B^2 - 4AC \rightarrow D$
: If D < 0
: Then
: Disp "NO SOLUTION"
: Else
: Disp "$B^2 - 4AC = $"
: Disp D
: Pause
: Disp "$\sqrt{(B^2 - 4AC)} = $"
: Disp $\sqrt{D}$
: End

SHARP EL-9600c

DISC
Input A
Input B
Input C
$B^2 - 4AC \Rightarrow D$
If D < 0 Goto 1
If D ≥ 0 Goto 2
Label 1
Print "NO SOLUTION"
End
Label 2
Print "$B^2 - 4AC = $
Print D
Wait
Print "$\sqrt{(B^2 - 4AC)} = $
Print $\sqrt{D}$
End

CASIO CFX-9850GA PLUS

DISC
"A = "? $\rightarrow$ A ↵
"B = "? $\rightarrow$ B ↵
"C = "? $\rightarrow$ C ↵
$B^2 - 4AC \rightarrow D$ ↵
If D < 0 ↵
Then "NO SOLUTION" ↵
Else "$B^2 - 4AC = $" ↵
D ◢

"$\sqrt{(B^2 - 4AC)} = $" ↵

$\sqrt{D}$ ↵

IfEnd ↵

2. To tell whether a trinomial can be factored, run the program on your calculator.
Enter the values of a, b, and c at the prompt.

Exercises

Use your program to determine whether the trinomial can be factored with integer coefficients.

1. $3t^2 - 8t + 40 = 0$

2. $11x^2 + 84x - 32 = 0$

3. $-7y^2 - 6y + 34 = 0$

4. $2x^2 + 5x - 18 = 0$

5. $9w^2 - 7w - 12 = 0$

6. $-b^2 + 14b - 120 = 0$

NAME _____ DATE _____

Practice A

For use with pages 611–617

Use the model to write the factors of the trinomial.

1.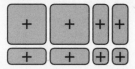

2.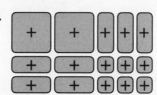

Match the trinomial with a correct factorization.

3. $2x^2 - 11x - 6$ **A.** $(2x + 3)(x + 2)$

4. $2x^2 + 11x - 6$ **B.** $(2x - 1)(x + 6)$

5. $2x^2 - 7x + 6$ **C.** $(2x + 1)(x - 6)$

6. $2x^2 + 7x + 6$ **D.** $(2x - 3)(x - 2)$

Choose the correct factorization. If neither is correct, find the correct factorization.

7. $2x^2 - 3x - 20$

   A. $(2x + 5)(x - 4)$

   B. $(2x + 10)(x - 2)$

8. $3x^2 + 11x - 4$

   A. $(3x + 1)(x - 4)$

   B. $(3x - 2)(x + 2)$

9. $3x^2 - 12x + 12$

   A. $(x - 3)(3x - 4)$

   B. $(x - 2)(3x - 6)$

Factor the trinomial if possible. If it cannot be factored, write *not factorable.*

10. $2x^2 - 5x - 3$ **11.** $3x^2 + 10x - 8$ **12.** $7x^2 - 31x + 12$

13. $3x^2 + 8x - 5$ **14.** $5x^2 + 7x + 2$ **15.** $6x^2 - 11x + 3$

16. $30x^2 + x - 1$ **17.** $5x^2 - 7x + 3$ **18.** $2x^2 - 9x - 5$

Solve the equation by factoring.

19. $3x^2 + 9x - 12 = 0$ **20.** $3x^2 + 13x - 10 = 0$ **21.** $2x^2 + 3x - 5 = 0$

22. $5x^2 - 8x + 3 = 0$ **23.** $3x^2 + 14x + 15 = 0$ **24.** $8x^2 - 16x + 6 = 0$

25. $7x^2 + 11x - 30 = 0$ **26.** $5x^2 - 22x - 15 = 0$ **27.** $2x^2 - 15x + 28 = 0$

28. *Ball Toss*   A ball is tossed into the air from a height of 10 feet with an initial velocity of 12 feet per second. Find the time t (in seconds) for the object to reach the ground by solving the equation $-16t^2 + 12t + 10 = 0$.

Practice B

For use with pages 611–617

Match the trinomial with a correct factorization.

1. $2x^2 + 2x - 12$ **A.** $(2x + 3)(x + 4)$

2. $2x^2 + 14x + 12$ **B.** $2(x - 2)(x + 3)$

3. $2x^2 - 10x + 12$ **C.** $2(x + 1)(x + 6)$

4. $2x^2 - 2x - 12$ **D.** $2(x - 1)(x - 6)$

5. $2x^2 + 11x + 12$ **E.** $2(x + 2)(x - 3)$

6. $2x^2 - 14x + 12$ **F.** $2(x - 2)(x - 3)$

Choose the correct factorization. If neither is correct, find the correct factorization.

7. $2x^2 + 4x - 16$

    A. $(2x + 4)(x - 4)$

    B. $(2x + 8)(x - 2)$

8. $5x^2 - 17x + 6$

    A. $(5x + 1)(x + 6)$

    B. $(5x - 3)(x - 2)$

9. $6x^2 - 17x + 5$

    A. $(3x - 1)(2x - 5)$

    B. $(3x + 1)(2x + 5)$

Factor the trinomial if possible. If it cannot be factored, write *not factorable*.

10. $2x^2 + 11x + 15$ **11.** $3x^2 + 10x - 7$ **12.** $10x^2 + 13x - 3$

13. $10x^2 + 17x + 3$ **14.** $8x^2 + 2x - 3$ **15.** $3x^2 + 2x - 2$

16. $12x^2 + 16x - 3$ **17.** $4x^2 - 3x + 8$ **18.** $10x^2 - 9x - 9$

Solve the equation by factoring.

19. $6x^2 - 10x - 4 = 0$ **20.** $6x^2 - 27x + 27 = 0$ **21.** $3x^2 + 5x + 2 = 0$

22. $8x^2 + 10x + 3 = 0$ **23.** $4x^2 - 8x - 5 = 0$ **24.** $12x^2 - 5x - 3 = 0$

25. $15x^2 + 16x - 15 = 0$ **26.** $8x^2 - 22x + 5 = 0$ **27.** $6x^2 + 5x + 1 = 0$

28. *Summer Business*   Your friend's weekly revenue R (in dollars) from her tie-dye T-shirt business can be modeled by

$$R = -2t^2 + 37t + 60$$

where t represents the week of sales, with $t = 0$ for the first week. In the first week, 3 T-shirts were sold. After that, the sales increased by 2 T-shirts per week. Did the price of T-shirts remain constant during the 8-week summer season? Explain.

29. *Cliff Diving*   A cliff diver jumps from a ledge 96 feet above the ocean with an initial upward velocity of 16 feet per second. How long will it take until the diver enters the water?

Choose the correct factorization. If neither is correct, find the correct factorization.

1. $6x^2 + 5x - 4$
   A. $(3x + 4)(2x - 1)$
   B. $(3x - 4)(2x + 1)$

2. $6x^2 - 13x - 5$
   A. $(6x - 6)(x + 1)$
   B. $(6x + 6)(x - 1)$

3. $12x^2 + 7x - 12$
   A. $(4x + 3)(3x - 4)$
   B. $(4x - 3)(3x + 4)$

Factor the trinomial if possible. If it cannot be factored, write *not factorable*.

4. $2x^2 - x - 21$

5. $3x^2 + 9x - 7$

6. $9x^2 + 6x + 1$

7. $3x^2 + 11x + 10$

8. $2x^2 - x - 6$

9. $3x^2 + x - 1$

10. $14x^2 - 19x - 40$

11. $4x^2 - 3x + 7$

12. $6x^2 - 36x + 54$

Solve the equation by factoring.

13. $2x^2 + 7x + 3 = 0$

14. $3x^2 + 14x - 5 = 0$

15. $3x^2 + 11x - 4 = 0$

16. $6x^2 + 13x + 5 = 0$

17. $3x^2 + 7x = -2$

18. $12x^2 = 5x + 3$

19. $10x^2 + 5 = -15x$

20. $12x^2 + 32x = -5$

21. $140x^2 + 300x = -40x - 120$

Solve the equation by factoring, by square roots, or by using the quadratic formula.

22. $4x^2 - 9 = 0$

23. $x^2 + 6x = 0$

24. $x^2 - 4x + 1 = 0$

25. $x^2 + 21 = 10x$

26. $2x^2 + 12x + 10 = -8$

27. $12x^2 + x - 1 = 0$

28. $2x^2 + 3x + 5 = 8$

29. $4x^2 - 64 = 0$

30. $18x^2 - 27x = 35$

Vertical Motion **In Exercises 31 and 32, use the vertical motion model** $h = -16t^2 + vt + s$, **where *h* is the height (in feet), *t* is the time in motion (in seconds), *v* is the initial velocity (in feet per second), and *s* is the initial height (in feet). Solve by factoring.**

31. A baseball player releases a baseball at a height of 6 feet with an initial velocity of 46 feet per second. Find the time (in seconds) for the ball to reach the ground.

32. A miniature rocket is launched off a roof 25 feet above the ground with an initial velocity of 30 feet per second. How much time will elapse before the rocket reaches the ground?

Reteaching with Practice

For use with pages 611–617

GOAL **Factor a quadratic expression of the form $ax^2 + bx + c$ and solve quadratic equations by factoring**

VOCABULARY

To factor quadratic polynomials whose leading coefficient is not 1, find the factors of a (m and n) and the factors of c (p and q) so that the sum of the outer and inner products (mq and pn) is b.

$$ax^2 + bx + c = (mx + p)(nx + q) \qquad b = mq + pn$$

with $c = pq$ and $a = mn$

EXAMPLE 1 *One Pair of Factors for a and c*

Factor $3x^2 + 7x + 2$.

SOLUTION

Test the possible factors of a (1 and 3) and c (1 and 2).

Try $a = 1 \cdot 3$ and $c = 1 \cdot 2$.

$\qquad (1x + 1)(3x + 2) = 3x^2 + 5x + 2 \qquad$ Not correct

Try $a = 1 \cdot 3$ and $c = 2 \cdot 1$.

$\qquad (1x + 2)(3x + 1) = 3x^2 + 7x + 2 \qquad$ Correct

The correct factorization of $3x^2 + 7x + 2$ is $(x + 2)(3x + 1)$.

Exercises for Example 1

Factor the trinomial.

1. $5x^2 + 11x + 2$ **2.** $2x^2 + 5x + 3$ **3.** $3x^2 + 10x + 7$

EXAMPLE 2 *Several Pairs of Factors for a and c*

Factor $4x^2 - 13x + 10$.

SOLUTION

Both factors of c must be negative, because b is negative and c is positive.

Reteaching with Practice

For use with pages 611–617

Test the possible factors of a and c.

FACTORS OF a and c	PRODUCT	CORRECT?
$a = 1 \cdot 4$ and $c = (-1)(-10)$	$(x - 1)(4x - 10) = 4x^2 - 14x + 10$	No
$a = 1 \cdot 4$ and $c = (-10)(-1)$	$(x - 10)(4x - 1) = 4x^2 - 41x + 10$	No
$a = 1 \cdot 4$ and $c = (-2)(-5)$	$(x - 2)(4x - 5) = 4x^2 - 13x + 10$	Yes
$a = 1 \cdot 4$ and $c = (-5)(-2)$	$(x - 5)(4x - 2) = 4x^2 - 22x + 10$	No
$a = 2 \cdot 2$ and $c = (-1)(-10)$	$(2x - 1)(2x - 10) = 4x^2 - 22x + 10$	No
$a = 2 \cdot 2$ and $c = (-10)(-1)$	$(2x - 10)(2x - 1) = 4x^2 - 22x + 10$	No
$a = 2 \cdot 2$ and $c = (-2)(-5)$	$(2x - 2)(2x - 5) = 4x^2 - 14x + 10$	No
$a = 2 \cdot 2$ and $c = (-5)(-2)$	$(2x - 5)(2x - 2) = 4x^2 - 14x + 10$	No

The correct factorization of $4x^2 - 13x + 10$ is $(x - 2)(4x - 5)$.

Exercises for Example 2

Factor the trinomial.

4. $9x^2 + 65x + 14$ **5.** $6x^2 - 23x + 15$ **6.** $8x^2 + 38x + 9$

EXAMPLE 3 *Solving a Quadratic Equation*

Solve the equation $3x^2 - x = 10$ by factoring.

SOLUTION

$3x^2 - x = 10$	Write equation.
$3x^2 - x - 10 = 0$	Write in standard form.
$(3x + 5)(x - 2) = 0$	Factor left side.
$(3x + 5) = 0$ or $(x - 2) = 0$	Use zero-product property.
$3x + 5 = 0$	Set first factor equal to 0.
$x = -\frac{5}{3}$	Solve for x.
$x - 2 = 0$	Set second factor equal to 0.
$x = 2$	Solve for x.

The solutions are $-\frac{5}{3}$ and 2.

Exercises for Example 3

Solve the equation by factoring.

7. $2x^2 + 7x + 3 = 0$ **8.** $5n^2 - 17n = -6$ **9.** $6x^2 - x - 2 = 0$

Lesson 10.6

NAME _____ DATE _____

Quick Catch-Up for Absent Students

For use with pages 610–618

The items checked below were covered in class on (date missed) _____

Activity 10.6: Modeling the Factorization of $ax^2 + bx + c$ (p. 610)

____ **Goal:** Model the factorization of a trinomial of the form $ax^2 + bx + c$ using algebra tiles.

Lesson 10.6: Factoring $ax^2 + bx + c$

____ **Goal 1:** Factor a quadratic expression of the form $ax^2 + bx + c$. (pp. 611–612)

Material Covered:

____ Example 1: One Pair of Factors for a and c

____ Example 2: One Pair of Factors for a and c

____ Example 3: Several Pairs of Factors for a and c

____ Example 4: A Common Factor for a, b, and c

____ Student Help: Study Tip

____ **Goal 2:** Solve quadratic equations by factoring. (p. 613)

Material Covered:

____ Example 5: Solving a Quadratic Equation

____ Student Help: Look Back

____ Example 6: Writing a Quadratic Model

Activity 10.6: Programming the Discriminant (p. 618)

____ **Goal:** Program a graphing calculator or a computer to find the value of the discriminant of $ax^2 + bx + c$.

____ Student Help: Keystroke Help

____ Other (specify) _____

_____

Homework and Additional Learning Support

____ Textbook (specify) pp. 614–617_____

_____

____ *Reteaching with Practice* worksheet (specify exercises)_____

____ *Personal Student Tutor* for Lesson 10.6

Interdisciplinary Application

For use with pages 611–617

The Art of Africa

ART   The letters of the alphabet from *A* to *Z* (excluding *W* and *X*) are represented by nonzero integers from −12 to 12. Copy and complete the table by factoring the polynomials in Exercises 1-6. Then use the table to match the coded words in Exercises 7-10 with a piece of African art.

Letter	A	B	C	D	E	F	G	H	I	J	K	L
Code Number	1				9				2			
Letter	M	N	O	P	Q	R	S	T	U	V	Y	Z
Code Number	3				6				11			

1. $8x^2 - 35x + 12 = (Ax + B)(Cx + D)$

2. $90x^2 - 143x + 56 = (Ex + F)(Gx + H)$

3. $10x^2 - 57x + 54 = (Ix + J)(Kx + L)$

4. $12x^2 - 43x + 10 = (Mx + N)(Ox + P)$

5. $42x^2 - 89x + 22 = (Qx + R)(Sx + T)$

6. $132x^2 - 199x + 60 = (Ux + V)(Yx + Z)$

7.

−1	1	−2	−2	9	−11	−10	9	−3		−12	9	7	7	9	−6

8.

8	1	−11	−12	9	−3		7	−2	4	4	−6

9.

3	1	7	1	2		−10	9	8	5	−6	1	8	9

10.

2	−12	4	−11	12		3	1	7	5

Challenge: Skills and Applications

For use with pages 611–617

In Exercises 1–4, factor the trinomial.

1. $x^2 - 9xy + 14y^2$

2. $6a^2 - 5ab - 4b^2$

3. $12p^2 + 17pq - 5q^2$

4. $2a^2b^2 - 3ab - 20$

In Exercises 5 and 6, use substitution to factor the trinomial.

Example: $6(x - 4)^2 + 11(x - 4) - 35$

Solution: Let $y = x - 4$. Then the polynomial becomes $6y^2 + 11y - 35$.

$$6y^2 + 11y - 35 = (3y - 5)(2y + 7)$$

Substituting back, $[3(x - 4) - 5][2(x - 4) + 7] = (3x - 17)(2x - 1)$

5. $4(x + 5)^2 - 33(x + 5) - 27$

6. $15x + \sqrt{x} - 2$

In Exercises 7–10, use substitution to factor and then solve the equation.

7. $n^6 - 7n^3 - 8 = 0$

8. $4y^4 - 13y^2 + 9 = 0$

9. $2(x - 3)^2 + (x - 3) - 10 = 0$

10. $\dfrac{6}{t^2} - \dfrac{17}{t} + 5 = 0$

In Exercises 11–13, use the following information.

Kesia Wilson wants to send some delicate lab specimens through the mail in a cubical box. To ensure the safe arrival of the contents, she wants to pack the box in another cube whose edges are 2 inches longer than the edges of the first box, and she wants to fill the space between the two cubes with padding. She has 296 cubic inches of padding. Kesia wants to know how long an edge of the first box should be in order that this amount of padding will exactly fill the space between the two cubes.

11. Write a polynomial model to fit the situation where x is the length of an edge of the smaller cube.

12. Simplify the equation from Exercise 11 to a quadratic in standard form.

13. How long should an edge of the first box be in order that Kesia's padding will exactly fill the space between the two cubes?

NAME _____ DATE _____

Quiz 2

For use after Lessons 10.4–10.6

1. Solve the equation $(x + 2)(x - 6) = 0$. *(Lesson 10.4)*

2. Name the *x*-intercepts and the vertex of the graph of
$y = (x + 3)(x - 1)$. Then sketch the graph of the function.
(Lesson 10.4)

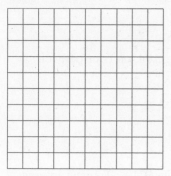

3. Factor the trinomial $x^2 - 7x + 10$. *(Lesson 10.5)*

4. Solve $x^2 - x = 20$ by factoring. *(Lesson 10.5)*

5. Factor the trinomial $4x^2 + 5x - 9$. *(Lesson 10.6)*

6. Solve $5x^2 + 8x - 21 = 0$ by factoring. *(Lesson 10.6)*

Answers

1. _____

2. _____

Use grid at left.

3. _____

4. _____

5. _____

6. _____

TEACHER'S NAME _____ CLASS _____ ROOM _____ DATE _____

Lesson Plan

2-day lesson (See *Pacing the Chapter,* TE pages 572C–572D)   **For use with pages 619–624**

 GOALS 1. **Use special product patterns to factor quadratic polynomials.**
          2. **Solve quadratic equations by factoring.**

State/Local Objectives _____

_____

✓ Check the items you wish to use for this lesson.

STARTING OPTIONS
____ Homework Check: TE page 614; Answer Transparencies
____ Warm-Up or Daily Homework Quiz: TE pages 619 and 616, CRB page 94, or Transparencies

TEACHING OPTIONS
____ Motivating the Lesson: TE page 620
____ Lesson Opener (Activity): CRB page 95 or Transparencies
____ Graphing Calculator Activity with Keystrokes: CRB pages 96–98
____ Examples: Day 1: 1–5, SE pages 619–621; Day 2: 6, SE page 621
____ Extra Examples: Day 1: TE pages 620–621 or Transp.; Day 2: TE page 621 or Transp.; Internet
____ Closure Question: TE page 621
____ Guided Practice: SE page 622; Day 1: Exs. 1–17; Day 2: none

APPLY/HOMEWORK
Homework Assignment
____ Basic   Day 1: 18–62 even, 63, 64; Day 2: 19–61 odd, 65, 66, 70–72, 75, 80, 85, 90, 95
____ Average   Day 1: 18–62 even, 63, 64; Day 2: 19–61 odd, 65–72, 75, 80, 85, 90, 95
____ Advanced   Day 1: 18–62 even, 63, 64; Day 2: 19–61 odd, 65–73, 75, 80, 85, 90, 95

Reteaching the Lesson
____ Practice Masters: CRB pages 99–101 (Level A, Level B, Level C)
____ Reteaching with Practice: CRB pages 102–103 or Practice Workbook with Examples
____ Personal Student Tutor

Extending the Lesson
____ Applications (Real-Life): CRB page 105
____ Challenge: SE page 624; CRB page 106 or Internet

ASSESSMENT OPTIONS
____ Checkpoint Exercises: Day 1: TE pages 620–621 or Transp.; Day 2: TE page 621 or Transp.
____ Daily Homework Quiz (10.7): TE page 624, CRB page 109, or Transparencies
____ Standardized Test Practice: SE page 624; TE page 624; STP Workbook; Transparencies

Notes _____

_____

_____

TEACHER'S NAME _____ CLASS _____ ROOM _____ DATE _____

Lesson Plan for Block Scheduling
1-day lesson (See *Pacing the Chapter*, TE pages 572C–572D)   For use with pages 619–624

GOALS   1. Use special product patterns to factor quadratic polynomials.
2. Solve quadratic equations by factoring.

State/Local Objectives _____

_____

_____

✓ Check the items you wish to use for this lesson.

STARTING OPTIONS
____ Homework Check: TE page 614; Answer Transparencies
____ Warm-Up or Daily Homework Quiz: TE pages 619 and
     616, CRB page 94, or Transparencies

TEACHING OPTIONS
____ Motivating the Lesson: TE page 620
____ Lesson Opener (Activity): CRB page 95 or Transparencies
____ Graphing Calculator Activity with Keystrokes: CRB pages 96–98
____ Examples 1–6: SE pages 619–621
____ Extra Examples: TE pages 620–621 or Transparencies; Internet
____ Closure Question: TE page 621
____ Guided Practice Exercises: SE page 622

APPLY/HOMEWORK
Homework Assignment
____ Block Schedule: 18–72, 75, 80, 85, 90, 95

Reteaching the Lesson
____ Practice Masters: CRB pages 99–101 (Level A, Level B, Level C)
____ Reteaching with Practice: CRB pages 102–103 or Practice Workbook with Examples
____ Personal Student Tutor

Extending the Lesson
____ Applications (Real-Life): CRB page 105
____ Challenge: SE page 624; CRB page 106 or Internet

ASSESSMENT OPTIONS
____ Checkpoint Exercises: TE pages 620–621 or Transparencies
____ Daily Homework Quiz (10.7): TE page 624, CRB page 109, or Transparencies
____ Standardized Test Practice: SE page 624; TE page 624; STP Workbook; Transparencies

Notes _____

_____

_____

_____

CHAPTER PACING GUIDE	
Day	**Lesson**
1	Assess Ch. 9; 10.1 (all)
2	10.2 (all); 10.3 (begin)
3	10.3 (end); 10.4 (all)
4	10.5 (all)
5	10.6 (all)
6	**10.7 (all)**
7	10.8 (all)
8	Review/Assess Ch. 10

Lesson 10.7

NAME _____ DATE _____

WARM-UP EXERCISES

For use before Lesson 10.7, pages 619–624

Find the product.

1. $(2n + 3)^2$

2. $(p - 6)^2$

3. $(2y - 1)(2y + 1)$

Factor the trinomial.

4. $x^2 - 4x - 5$

5. $x^2 + 2x + 1$

···

DAILY HOMEWORK QUIZ

For use after Lesson 10.6, pages 610–618

Factor, if possible. If not, write *not factorable*.

1. $14x^2 + 36x - 18$

2. $x^2 + 4x - 7$

3. $4n^2 + 4n - 288$

Solve the equation.

4. $6x^2 + 19x - 36 = 0$

5. $5y^2 + 22y = 15$

6. $0.4z^2 - 0.6 = z$

Algebra 1
Chapter 10 Resource Book

LESSON

10.7

NAME ———————————————————— DATE —————

Activity Lesson Opener

For use with pages 619–624

Available as
a transparency

SET UP: Work with a partner.

Find each product.

1. $(x + 1)(x - 1)$
2. $(m + 2)(m - 2)$
3. $(y + 4)(y - 4)$
4. $(n + 8)(n - 8)$

Use Questions 1–4 to answer Questions 5–7.

5. Describe the similarities among the answers.

6. Compare the first term of each answer to the first term of each of the binomials in the corresponding question. What do you notice?

7. Compare the last term of each answer to the second term of each of the binomials in the corresponding question. What do you notice?

Find each product.

8. $(x + 1)^2$
9. $(m + 2)^2$
10. $(y - 4)^2$
11. $(n - 8)^2$

Use Questions 8–11 to answer Questions 12–14.

12. Describe the similarities among the answers.

13. Compare the first term of each answer to the first term of the binomial in the corresponding question. What do you notice?

14. Compare the last term of each answer to the second term of the binomial in the corresponding question. What do you notice?

15. Compare the middle term of each answer to twice the product of the two terms of the binomial in the corresponding question. What do you notice?

NAME _____ DATE _____

Graphing Calculator Activity

For use with pages 619–624

GOAL **To check solutions of quadratic equations**

In Lesson 10.7, you will learn to check your solutions to quadratic
equations using the *Intersect* feature of your graphing calculator. An
alternate method for checking your solutions is to use your graphing
calculator's *Table* feature.

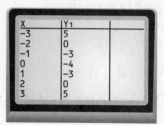

Activity

❶ Earlier in this chapter, you learned how to solve quadratic equations by
factoring. Solve each quadratic equation.

    a. $x^2 - 5x - 14 = 0$ **b.** $6x^2 - 6x - 72 = 0$ **c.** $4x^2 + 24x + 36 = 0$

❷ Enter the left-hand side of the equation in part (a) of Step 1 into your
graphing calculator as equation Y_1.

❸ Use the *Table* feature of your graphing calculator to find the x-value(s) for which the
y-value equals 0. A y-value of 0 indicates a solution of the equation. Scroll the table
up and down to be sure you have found all of the solutions. Did you solve the
equation correctly in part (a) of Step 1?

❹ Repeat Steps 2 and 3 for parts (b) and (c) of Step 1.

❺ What is different about the solution(s) to part (c) comparred to parts (a) and
(b)? Why?

Exercises

**In Exercises 1–6, solve the equation. Then use the *Table* feature of your
graphing calculator to check your solution(s).**

1. $x^2 + 2x + 1 = 0$ **2.** $x^2 - 7x + 10 = 0$ **3.** $2x^2 - 10x + 12 = 0$

4. $4x^2 + 4x - 24 = 0$ **5.** $25x^2 + 50x + 25 = 0$ **6.** $9x^2 - 54x + 81 = 0$

7. Consider the equations in Exercises 1–6 that have only one solution. Can
you describe a way to predict if a quadratic equation will have only one
solution?

See page 97 for keystrokes.

NAME _____ DATE _____

Graphing Calculator Activity

For use with pages 619–624

TI-82

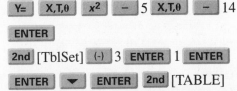

[Y=] [X,T,θ] [x²] [−] 5 [X,T,θ] [−] 14
[ENTER]
[2nd] [TblSet] [(-)] 3 [ENTER] 1 [ENTER]
[ENTER] [▼] [ENTER] [2nd] [TABLE]

Note: You can scroll the table up by pressing [▲] or down by pressing [▼].

[Y=] [CLEAR] 6 [X,T,θ] [x²] [−] 6 [X,T,θ] [−] 72
[ENTER]
[2nd] [TblSet] [(-)] 3 [ENTER] [2nd]
[TABLE]

[Y=] [CLEAR] 4 [X,T,θ] [x²] [+] 24 [X,T,θ] [+]
36 [ENTER]
[2nd] [TblSet] [(-)] 3 [ENTER] [2nd]
[TABLE]

TI-83

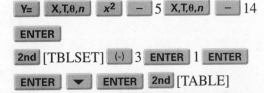

[Y=] [X,T,θ,n] [x²] [−] 5 [X,T,θ,n] [−] 14
[ENTER]
[2nd] [TBLSET] [(-)] 3 [ENTER] 1 [ENTER]
[ENTER] [▼] [ENTER] [2nd] [TABLE]

Note: You can scroll the table up by pressing [▲] or down by pressing [▼].

[Y=] [CLEAR] 6 [X,T,θ,n] [x²] [−] 6
[X,T,θ,n] [−] 72 [ENTER]
[2nd] [TBLSET] [(-)] 3 [ENTER] [2nd]
[TABLE]

[Y=] [CLEAR] 4 [X,T,θ,n] [x²] [+] 24
[X,T,θ,n] [+] 36 [ENTER]
[2nd] [TBLSET] [(-)] 3 [ENTER] [2nd]
[TABLE]

Sharp EL-9600c

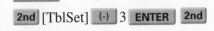
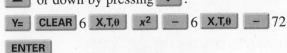

[Y=] [X/θ/T/n] [x²] [−] 5 [X/θ/T/n] [−] 14
[ENTER]
[2ndF] [TBLSET] [ENTER] [▼] [(-)] 3
[ENTER] 1 [ENTER] [TABLE]

Note: You can scroll the table up by pressing [▲] or down by pressing [▼].

[Y=] [CL] 6 [X/θ/T/n] [x²] [−] 6 [X/θ/T/n] [−]
72 [ENTER]
[2ndF] [TBLSET] [▼] [(-)] 3 [ENTER]
[TABLE]

[Y=] [CL] 4 [X/θ/T/n] [x²] [+] 24 [X/θ/T/n] [+]
36 [ENTER]
[2ndF] [TBLSET] [▼] [(-)] 3 [ENTER]
[TABLE]

Casio CFX-9850Ga PLUS

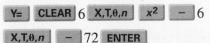

From the main menu, choose TABLE.

[X,θ,T] [x²] [−] 5 [X,θ,T] [−] 14 [EXE]
[F5] [(-)] 10 [EXE] 10 [EXE] 1 [EXE]
[EXIT] [F6]

Note: You can scroll the table up by pressing [▲] or down by pressing [▼].

[EXIT] [▲] 6 [X,θ,T] [x²] [−] 6 [X,θ,T] [−] 72
[EXE]
[F6]

[EXIT] [▲] 4 [X,θ,T] [x²] [+] 24 [X,θ,T] [+]
36 [EXE]
[F6]

NAME _____ DATE _____

Graphing Calculator Activity Keystrokes

For use with page 620.

Keystrokes for Example 3

TI-82

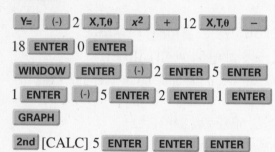

TI-83

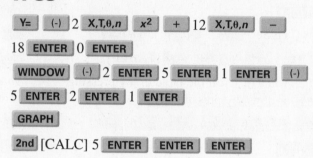

SHARP EL-9600c

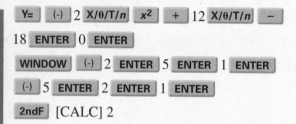

CASIO CFX-9850GA PLUS

From the main menu, choose GRAPH.

| (-) | 2 | X,θ,T | x² | + | 12 | X,θ,T | − | 18 | EXE | 0 | EXE |

| SHIFT | F3 | (-) | 2 | EXE | 5 | EXE | 1 | EXE | (-) | 5 | EXE |

2 EXE 1 EXE EXIT F6 F5 F5

Algebra 1
Chapter 10 Resource Book

Lesson 10.7

NAME _____ DATE _____

Practice A

For use with pages 619–624

Factor out the greatest common monomial factor.

1. $3x^2 + 6x + 3$

2. $2x^2 - 8x + 8$

3. $4x^2 + 8x + 4$

4. $3x^2 - 12x + 12$

5. $5x^2 + 30x + 45$

6. $-3x^2 - 18x - 27$

Match the trinomial with a correct factorization.

7. $x^2 - 6x + 9$

8. $x^2 + 6x + 9$

9. $2x^2 - 12x + 18$

10. $3x^2 + 18x + 27$

11. $2x^2 - 18$

12. $3x^2 - 27$

A. $3(x + 3)^2$

B. $(x + 3)^2$

C. $(x - 3)^2$

D. $2(x - 3)(x + 3)$

E. $3(x + 3)(x - 3)$

F. $2(x - 3)^2$

Factor the expression.

13. $x^2 - 25$

14. $x^2 - 49$

15. $4x^2 - 9$

16. $x^2 - \frac{1}{4}$

17. $x^2 - y^2$

18. $25x^2 - 16$

19. $9x^2 - 16$

20. $16 - x^2$

21. $200x^2 - 18$

Factor the expression.

22. $x^2 + 10x + 25$

23. $x^2 - 18x + 81$

24. $4x^2 + 4x + 1$

25. $x^2 + x + \frac{1}{4}$

26. $x^2 - 2xy + y^2$

27. $4x^2 + 28x + 49$

28. $25x^2 - 20x + 4$

29. $9x^2 - 24xy + 16y^2$

30. $20x^2 + 60x + 45$

Use factoring to solve the equation. Use a graphing calculator to check your solution if you wish.

31. $2x^2 - 18 = 0$

32. $3x^2 + 12x + 12 = 0$

33. $x^2 - \frac{4}{9} = 0$

34. $10x^2 - 80x + 160 = 0$

35. $7x^2 - 28 = 0$

36. $4x^2 - 40x + 100 = 0$

37. *Washers* Washers are available in various sizes. Find an expression for the area of one flat side of a washer. Factor the expression. What is the area if $x = 5$ centimeters and $y = 2$ centimeters?

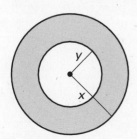

NAME _____ DATE _____

Practice B

For use with pages 619–624

Match the trinomial with a correct factorization.

1. $x^2 - 16$

2. $x^2 + 8x + 16$

3. $2x^2 + 16x + 32$

4. $2x^2 - 16x + 32$

5. $x^2 - 8x + 16$

6. $2x^2 - 32$

A. $2(x + 4)^2$

B. $(x + 4)^2$

C. $(x - 4)^2$

D. $2(x - 4)(x + 4)$

E. $(x + 4)(x - 4)$

F. $2(x - 4)^2$

Factor the expression.

7. $x^2 - 9$

8. $x^2 - 144$

9. $x^2 - \frac{1}{9}$

10. $x^2 - 0.16$

11. $4x^2 - 49$

12. $9x^2 - 25$

13. $p^2 - q^2$

14. $64x^2 - 9y^2$

15. $100 - x^2$

16. $49 - x^2$

17. $4a^2 - b^2$

18. $12x^2 - 48$

Factor the expression.

19. $x^2 + 18x + 81$

20. $x^2 - 8x + 16$

21. $4x^2 + 4x + 1$

22. $25x^2 - 30x + 9$

23. $a^2 - 2ab + b^2$

24. $x^2 + 4xy + 4y^2$

25. $4x^2 + 2x + \frac{1}{4}$

26. $9x^2 - 0.6x + 0.01$

27. $8x^2 + 8xy + 2y^2$

28. $3x^2 - 30xy + 75y^2$

29. $25 - 20x + 4x^2$

30. $36 + 48x + 16x^2$

Use factoring to solve the equation. Use a graphing calculator to check your solution if you wish.

31. $3x^2 - 48 = 0$

32. $4x^2 - 20x + 25 = 0$

33. $3x^2 - \frac{3}{4} = 0$

34. $9x^2 - 24x + 16 = 0$

35. $-5x^2 + 20 = 0$

36. $\frac{4}{3}x^2 + \frac{20}{3}x + \frac{25}{3} = 0$

37. *Quilt*   A square quilt for a child's bed has a border made up of 36 pieces with an area of x square inches each, and 4 small squares with an area of 1 square inch each. The main part of the quilt is made up of 81 squares with an area of x^2 square inches each. Find an expression for the area of the quilt. Factor the expression. If the quilt is 5 feet by 5 feet, what are the dimensions of the inside squares?

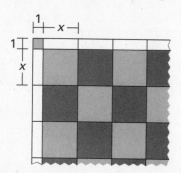

NAME _____ DATE _____

Practice C

For use with pages 619–624

Factor the expression.

1. $x^2 - 9$

2. $121 - x^2$

3. $9 - 4x^2$

4. $9x^2 - 49$

5. $25x^2 - 4y^2$

6. $49x^2 - \frac{1}{4}$

7. $\frac{1}{4}x^2 - \frac{1}{9}y^2$

8. $25x^2 - 0.04$

9. $9x^2 - 49y^2$

10. $18x^2 - 32$

11. $20x^2 - \frac{5}{4}$

12. $-7x^2 + 28$

13. $-27x^2 + 12y^2$

14. $x^4 - 16$

15. $4x^2 - 0.09$

Factor the expression.

16. $x^2 + 14x + 49$

17. $4x^2 + 4xy + y^2$

18. $9x^2 - 12x + 4$

19. $25 - 10x + x^2$

20. $9 + 12x + 4x^2$

21. $9x^2 - 12xy + 4y^2$

22. $x^2 + \frac{3}{2}x + \frac{9}{16}$

23. $4a^2 - 20ab + 25b^2$

24. $12x^2 + 84x + 147$

25. $-4x^2 + 24x - 36$

26. $25x^2 - 2x + 0.04$

27. $8x^2 + 48x + 72$

28. $-20x^2 - 60x - 45$

29. $49 - 42x + 9x^2$

30. $x^4 + 2x^2y^2 + y^4$

Use factoring to solve the equation. Use a graphing calculator to check your solution if you wish.

31. $18x^2 - 50 = 0$

32. $5x^2 - 40x + 80 = 0$

33. $7x^2 - \frac{7}{9} = 0$

34. $-36x^2 + 36x - 9 = 0$

35. $3x^2 - 108 = 0$

36. $\frac{4}{5}x^2 - \frac{12}{5}x + \frac{9}{5} = 0$

37. *Trapezoid* The formula for the area of a trapezoid is $A = \frac{1}{2}h(b_1 + b_2)$. Derive this formula by finding the sum of the areas of the rectangle and triangle that make up the trapezoid.

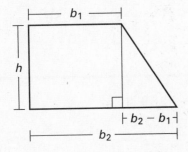

38. *Rectangle* Find an expression for the area of the shaded region. Factor the expression. If $x = 10$, what is the area of the rectangle?

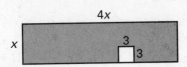

Algebra 1
Chapter 10 Resource Book

NAME _____ DATE _____

Reteaching with Practice

For use with pages 619–624

GOAL Use special product patterns to factor quadratic polynomials and solve quadratic equations by factoring

---

VOCABULARY

Factoring Special Products

Difference of Two Squares Pattern   **Example**

$a^2 - b^2 = (a + b)(a - b)$ $9x^2 - 16 = (3x + 4)(3x - 4)$

Perfect Square Trinomial Pattern   **Example**

$a^2 + 2ab + b^2 = (a + b)^2$ $x^2 + 8x + 16 = (x + 4)^2$

$a^2 - 2ab + b^2 = (a - b)^2$ $x^2 - 12x + 36 = (x - 6)^2$

EXAMPLE 1 *Factoring the Difference of Two Squares*

a. $n^2 - 25$

b. $4x^2 - y^2$

SOLUTION

a. $n^2 - 25 = n^2 - 5^2$ Write as $a^2 - b^2$.

$\qquad = (n + 5)(n - 5)$ Factor using difference of two squares pattern.

b. $4x^2 - y^2 = (2x)^2 - y^2$ Write as $a^2 - b^2$.

$\qquad = (2x + y)(2x - y)$ Factor using difference of two squares pattern.

Exercises for Example 1

Factor the expression.

1. $16 - 9y^2$

2. $4q^2 - 49$

3. $36 - 25x^2$

EXAMPLE 2 *Factoring Perfect Square Trinomials*

a. $x^2 - 6x + 9$

b. $9y^2 + 12y + 4$

SOLUTION

a. $x^2 - 6x + 9 = x^2 - 2(x)(3) + 3^2$ Write as $a^2 - 2ab + b^2$.

$\qquad = (x - 3)^2$ Factor using perfect square trinomials.

b. $9y^2 + 12y + 4 = (3y)^2 + 2(3y)(2) + 2^2$ Write as $a^2 + 2ab + b^2$.

$\qquad = (3y + 2)^2$ Factor using perfect square trinomial pattern.

Exercises for Example 2

Factor the expression.

4. $x^2 - 18x + 81$

5. $4n^2 + 20n + 25$

6. $16y^2 + 8y + 1$

Algebra 1
Chapter 10 Resource Book

NAME _____ DATE _____

Reteaching with Practice

For use with pages 619–624

EXAMPLE 3 *Solving a Quadratic Equation*

Solve the equation $2x^2 - 28x + 98 = 0$.

SOLUTION

$2x^2 - 28x + 98 = 0$	Write original equation.
$2(x^2 - 14x + 49) = 0$	Factor out common factor.
$2[x^2 - 2(7x) + 7^2] = 0$	Write as $a^2 - 2ab + b^2$.
$2(x - 7)^2 = 0$	Factor using perfect square trinomial pattern.
$x - 7 = 0$	Set repeated factor equal to 0.
$x = 7$	Solve for x.

The solution is 7.

Exercises for Example 3

Use factoring to solve the equation.

7. $x^2 - 20x + 100 = 0$ **8.** $4n^2 - 4n = -1$ **9.** $3z^2 - 24z + 48 = 0$

EXAMPLE 4 *Solving a Quadratic Equation*

Solve the equation $75 - 48x^2 = 0$.

SOLUTION

$75 - 48x^2 = 0$	Write original equation.
$3(25 - 16x^2) = 0$	Factor out common factor.
$3[5^2 - (4x)^2] = 0$	Write as $a^2 - b^2$.
$3(5 + 4x)(5 - 4x) = 0$	Factor using difference of two squares pattern.
$(5 + 4x) = 0$ or $(5 - 4x) = 0$	Use zero-product property.
$5 + 4x = 0$	Set first factor equal to 0.
$x = -\frac{5}{4}$	Solve for x.
$5 - 4x = 0$	Set second factor equal to 0.
$x = \frac{5}{4}$	Solve for x.

The solutions are $-\frac{5}{4}$ and $\frac{5}{4}$.

Exercises for Example 4

Use factoring to solve the equation.

10. $x^2 - 49 = 0$ **11.** $9y^2 - 64 = 0$ **12.** $4x^2 = 81$

Lesson 10.7

NAME _____ DATE _____

Quick Catch-Up for Absent Students

For use with pages 619–624

The items checked below were covered in class on (date missed) _____

Lesson 10.7: Factoring Special Products

_____ **Goal 1:** Use special product patterns to factor quadratic polynomials. (p. 619)

Material Covered:

_____ Example 1: Factoring the Difference of Two Squares

_____ Example 2: Factoring Perfect Square Trinomials

_____ **Goal 2:** Solve quadratic equations by factoring. (pp. 620–621)

Material Covered:

_____ Example 3: Graphical and Analytical Reasoning

_____ Student Help: Study Tip

_____ Example 4: Solving a Quadratic Equation

_____ Example 5: Writing and Using a Quadratic Model

_____ Example 6: Writing and Using a Quadratic Model

_____ Other (specify) _____

_____

Homework and Additional Learning Support

_____ Textbook (specify) _pp. 622–624_____

_____

_____ Internet: Extra Examples at www.mcdougallittell.com

_____ *Reteaching with Practice* worksheet (specify exercises)_____

_____ *Personal Student Tutor* for Lesson 10.7

NAME _____ DATE _____

Real-Life Application:
When Will I Ever Use This?

For use with pages 619–624

Manufacturing

You work for a company that manufactures round metal washers. The cost to produce round metal washers is directly related to the amount of metal used. And the amount of metal used is directly related to the area of the washer, as shown in the shaded region below.

In Exercises 1-4, use the diagram to the right.

1. Find the area of the large disc with radius R before the middle hole is stamped out. The area of a circle is $A = \pi \cdot r^2$, where A is the area and r is the radius of the circle.

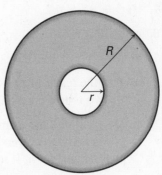

2. Find the area of the middle hole with radius r.

3. Write a formula for the area of the washer that has an outer radius R and a hole of radius r, using the formulas from parts a and b. Simplify the formula by factoring out all common monomial term(s).

4. You should recognize one of the factors in your formula as a "special product." Show your formula for the area of a washer in yet another form by rewriting the "special product" in your formula as two factors.

In Exercises 5-9, use the following information.

The process of making this washer starts with a square piece of metal and makes two circular stamps. The length of the square's side will be equal to the diameter of the washer (Diameter = twice the radius). The material outside the diameter of the washer as well as the hole from the middle becomes waste.

5. Write a formula for the area of the square piece of metal that will be used to stamp out the washer.

6. Create a formula for the amount of metal waste from the production of a single washer of outer radius R and inner radius r.

7. Using your formula from Exercise 3, find the amount of metal used to produce a washer with an outer radius of $\frac{3}{4}$ inch and an inner radius of $\frac{1}{2}$ inch.

8. Find the amount of waste in creating the same washer in Exercise 7.

9. Comparing the metal wasted to the area of the initial metal square, what percent of metal is wasted in creating one of these metal washers?

Lesson 10.7

Challenge: Skills and Applications

For use with pages 619–624

In Exercises 1–6, factor the expression. Tell which special product factoring pattern you used.

1. $49r^2s^2 - 100t^2$

2. $(b + 3)^2 - 49$

3. $(x + 8)^2 - 6(x + 8) + 9$

4. $r^4 - 18r^2 + 81$

5. $k^4 - 16$

6. $25 - (n^2 + 1)^2$

In Exercises 7–10, use factoring to solve the equation.

7. $16a^4 - 81 = 0$

8. $(a + 2)^2 - 49 = 0$

9. $x^4 - 12x^2 + 36 = 0$

10. $(x - 9)^2 + 10(x - 9) + 25 = 0$

11. a. Multiply $(x - 3)(x^3 + 3x^2 + 9x + 27)$

    b. Use the result of part (a) to suggest a formula for factoring a difference of two fourth powers. Use multiplication to check your result.

    c. Do you think the formula you found in part (b) can be extended to fifth powers, sixths powers, and so on? If so, state the formula for any positive integer n. If not, explain why not.

12. a. By rewriting $x^4 + 64$ in the form $(x^4 + 16x^2 + 64) - 16x^2$, write this expression as a difference of two squares.

    b. Use the formula for the difference of two squares to factor the expression you wrote in part (a).

    c. Use the method of parts (a) and (b) to factor $x^4 + 1024$.

TEACHER'S NAME _____ CLASS _____ ROOM _____ DATE _____

Lesson Plan

2-day lesson (See *Pacing the Chapter,* TE pages 572C–572D)   **For use with pages 625–632**

 GOALS   1. **Use the distributive property to factor a polynomial.**
                2. **Solve polynomial equations by factoring.**

State/Local Objectives _____

_____

✓ Check the items you wish to use for this lesson.

STARTING OPTIONS
____ Homework Check: TE page 622; Answer Transparencies
____ Warm-Up or Daily Homework Quiz: TE pages 625 and 624, CRB page 109, or Transparencies

TEACHING OPTIONS
____ Motivating the Lesson: TE page 626
____ Lesson Opener (Activity): CRB page 110 or Transparencies
____ Examples: Day 1: 1–7, SE pages 625–627; Day 2: 8–9, SE pages 627–628
____ Extra Examples: Day 1: TE pages 626–627 or Transp.; Day 2: TE pages 627–628 or Transp.
____ Closure Question: TE page 628
____ Guided Practice: SE page 629; Day 1: Exs. 1–14; Day 2: none

APPLY/HOMEWORK
Homework Assignment
____ Basic   Day 1: 16–50 even; Day 2: 21–51 odd, 52, 61–63, 70, 75, 80–82; Quiz 3: 1–25
____ Average   Day 1: 16–50 even; Day 2: 21–51 odd, 52, 55–57, 61–63, 70, 75, 80–82; Quiz 3: 1–25
____ Advanced   Day 1: 16–50 even; Day 2: 21–51 odd, 52–57, 61–68, 70, 75, 80–82; Quiz 3: 1–25

Reteaching the Lesson
____ Practice Masters: CRB pages 111–113 (Level A, Level B, Level C)
____ Reteaching with Practice: CRB pages 114–115 or Practice Workbook with Examples
____ Personal Student Tutor

Extending the Lesson
____ Applications (Real-Life): CRB page 117
____ Challenge: SE page 631; CRB page 118 or Internet

ASSESSMENT OPTIONS
____ Checkpoint Exercises: Day 1: TE pages 626–627 or Transp.; Day 2: TE pages 627–628 or Transp.
____ Daily Homework Quiz (10.8): TE page 631 or Transparencies
____ Standardized Test Practice: SE page 631; TE page 631; STP Workbook; Transparencies
____ Quiz (10.7–10.8): SE page 632

Notes _____

_____

_____

TEACHER'S NAME _____ CLASS _____ ROOM _____ DATE _____

Lesson Plan for Block Scheduling

1-day lesson (See *Pacing the Chapter,* TE pages 572C–572D)   For use with pages 625–632

GOALS   1. Use the distributive property to factor a polynomial.
            2. Solve polynomial equations by factoring.

State/Local Objectives _____

_____

_____

CHAPTER PACING GUIDE	
Day	**Lesson**
1	Assess Ch. 9; 10.1 (all)
2	10.2 (all); 10.3 (begin)
3	10.3 (end); 10.4 (all)
4	10.5 (all)
5	10.6 (all)
6	10.7 (all)
7	**10.8 (all)**
8	Review/Assess Ch. 10

✓ **Check the items you wish to use for this lesson.**

STARTING OPTIONS

____ Homework Check: TE page 622; Answer Transparencies

____ Warm-Up or Daily Homework Quiz: TE pages 625 and
       624, CRB page 109, or Transparencies

TEACHING OPTIONS

____ Motivating the Lesson: TE page 626

____ Lesson Opener (Activity): CRB page 110 or Transparencies

____ Examples 1–9: SE pages 625–628

____ Extra Examples: TE pages 626–628 or Transparencies; Internet

____ Closure Question: TE page 628

____ Guided Practice Exercises: SE page 629

APPLY/HOMEWORK

Homework Assignment

____ Block Schedule: 16, 18, 20–52, 55–57, 61–63, 70, 75, 80–82; Quiz 3: 1–25

Reteaching the Lesson

____ Practice Masters: CRB pages 111–113 (Level A, Level B, Level C)

____ Reteaching with Practice: CRB pages 114–115 or Practice Workbook with Examples

____ Personal Student Tutor

Extending the Lesson

____ Applications (Real-Life): CRB page 117

____ Challenge: SE page 631; CRB page 118 or Internet

ASSESSMENT OPTIONS

____ Checkpoint Exercises: TE pages 626–628 or Transparencies

____ Daily Homework Quiz (10.8): TE page 631 or Transparencies

____ Standardized Test Practice: SE page 631; TE page 631; STP Workbook; Transparencies

____ Quiz (10.7–10.8): SE page 632

Notes _____

_____

_____

Algebra 1
Chapter 10 Resource Book

NAME _____ DATE _____

WARM-UP EXERCISES

For use before Lesson 10.8, pages 625–632

Find the greatest common factor.

1. $9, 12$

2. $51, 85$

3. $30, 45, 105$

Use the distributive property to simplify.

4. $3x(4x + 9)$

5. $(-3)(2x^3 + 5)$

···

DAILY HOMEWORK QUIZ

For use after Lesson 10.7, pages 619–624

Factor the expression.

1. $36x^2 - 144$

2. $r^2 + 18r + 81$

3. $8y^2 - 40y + 50$

Solve the equation.

4. $4x^2 - 9 = 0$

5. $\frac{2}{3}x^2 + 8x + 24 = 0$

6. On his way to the basket, Michael Jordan jumped 3 ft in the air. How long did he stay in the air? Use the hang time model $h = 4t^2$.

Activity Lesson Opener

For use with pages 625–632

SET UP: Work with a partner.

Use the following information to complete the table.

Fill in the blanks. First complete the Across problem, then use that answer to help you complete the Down problem next to it. Then fill in the puzzle. Use only one digit or variable per box. If a variable is raised to an exponent, place the exponent in the same box as the variable.

ACROSS	*DOWN*
1. $2x(x + 3) =$ ____ $+ 6x$	**1.** $2x^2 + 6x =$ ____ $(x + 3)$
2. $4x(x^2 - 1) =$ ____ $- 4x$	**2.** $4x^3 - 4x =$ ____ $(x^2 - 1)$
3. $5x(2x + 1) = 10x^2 +$ ____	**3.** $10x^2 + 5x =$ ____ $(2x + 1)$
4. $x(3x - 1) = 3x^2 -$ ____	**4.** $3x^2 - x =$ ____ $(3x - 1)$
6. $3x^2(2 + x) =$ ____ $+ 3x^3$	**5.** $6x^2 + 3x^3 =$ ____ $(2 + x)$
7. $6x(x^2 - 2) = 6x^3 -$ ____	**6.** $6x^3 - 12x =$ ____ $(x^2 - 2)$
9. $20x(x + 4) = 20x^2 +$ ____	**8.** $20x^2 + 80x =$ ____ $(x + 4)$

Algebra 1
Chapter 10 Resource Book

Practice A

For use with pages 625–632

Find the greatest common factor and factor it out of the expression.

1. $3x + 18$

2. $-2c + 10$

3. $4y^2 + 4y + 8$

4. $6x^3 - 2x^2 + 2x$

5. $d^4 + d^3 - 2d^2$

6. $10a^3 - 12a^2 + 4a$

Tell whether the expression is factored completely. If the expression is not factored completely, write the complete factorization.

7. $2(x^2 + 1)$

8. $2(n^2 + 4n + 4)$

9. $3(x^2 - 1)$

10. $m(m^2 + 5m + 2)$

11. $2(x^2 + 5x + 6)$

12. $3t(t^2 - t + 10)$

Factor the expression completely.

13. $6(x + 1) + 7(x + 1)$

14. $c(c - 2) + 2(c - 2)$

15. $m(m + 3) - 5(m + 3)$

16. $2x(x + 4) + 7(x + 4)$

17. $14x - 28x^2$

18. $45mn - 30m^2$

19. $2x^2 + 16x + 14$

20. $5x^2 - 45$

21. $14t^2 - 35t - 21$

22. $x^3 + 6x^2 + 9x$

23. $x^3 + x^2 + 2x + 2$

24. $d^3 + d^2 + 3d + 3$

Solve the equation. Tell which solution method you used.

25. $x^2 + 4x + 3 = 0$

26. $x^2 - 16 = 0$

27. $9x^2 + 49 = 0$

28. $x^2 - 4x - 2 = 0$

29. $3x^2 + x + 1 = 0$

30. $4x^2 + 2x - 1 = 0$

31. *Area*   Find the area of the shaded region shown at the right. The area of the rectangle is $5x^2 + 12x + 10$ and the area of the circle is $x^2 + 2$. Write the area in factored form.

32. *Rectangle Area*   Find the dimensions of a rectangle if its area is $xy - 4x + 2y - 8$.

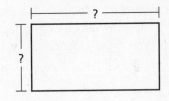

Practice B

For use with pages 625–632

Find the greatest common factor and factor it out of the expression.

1. $6x^2 + 10x$

2. $5c^3 - 25c^2 + 10c$

3. $15y^3 + 6y^2 - 21y$

4. $10x^4 + 16x^3 + 4x^2$

5. $4d^4 + d^3 - 3d^2$

6. $8a^5 - 10a^3 + 18a^2$

Tell whether the expression is factored completely. If the expression is not factored completely, write the complete factorization.

7. $3(x^2 + 9)$

8. $5(n^2 + 8n + 16)$

9. $2(x^2 - 4)$

10. $3m(m^2 + 9m + 27)$

11. $2(x^2 + 7x + 6)$

12. $3t(t^2 - 5t + 10)$

Factor the expression completely.

13. $6x^3 + 18x^2$

14. $3c^3 - 12c$

15. $-10m^3 - 2m$

16. $35a^3 - 28a^2$

17. $32x - 48x^2$

18. $35xy - 60x^2$

19. $3m^2 + 24m + 36$

20. $4x^2 + 4x - 80$

21. $2t^3 + 2t^2 - 12t$

22. $6x^3 + 24x^2 + 24x$

23. $x^3 + x^2 + 4x + 4$

24. $d^3 + 2d^2 + 3d + 6$

Solve the equation. Tell which solution method you used.

25. $x^2 + 7x + 6 = 0$

26. $x^2 - 5x + 9 = 0$

27. $4x^2 - 28x + 49 = 0$

28. $3x^2 - 6x + 2 = 0$

29. $7x^2 - 2x + 5 = 0$

30. $5x^2 + 4x - 3 = 0$

Vertical Motion   In Exercises 31 and 32, use the vertical motion model $h = 16t^2 - vt$, where h is the initial height (in feet), v is the initial velocity (in feet per second), and t is the time (in seconds) the object spends in the air.

31. *Baseball*   You toss a baseball from a height of 32 feet with an initial upward velocity of 16 feet per second. How long will it take the baseball to reach the ground?

32. *Rocket*   You launch a rocket from a height of 64 feet with an initial upward velocity of 48 feet per second. How long will it take the rocket to reach the ground?

Practice C

For use with pages 625–632

Find the greatest common factor and factor it out of the expression.

1. $3x^2 - 12x$

2. $4c^3 - 12c^2 + 8c$

3. $-7y^3 + 35y^2 - 7y$

4. $\frac{10}{3}x^3 + \frac{5}{3}x^2 + 35x$

5. $15d^4 - 6d^3 + 3d^2$

6. $8a^4b + 48a^2b - 88ab$

Tell whether the expression is factored completely. If the expression is not factored completely, write the complete factorization.

7. $3x(x^2 + 5)$

8. $2n(2n^2 - 9n - 5)$

9. $7x(9x^2 - 25)$

10. $6m(m^3 + 6m + 5)$

11. $8(6x^2 - 2x - 28)$

12. $-4t(5t^2 - 2t + 6)$

Factor the expression completely.

13. $21x^2 - 15x$

14. $-4c^3 + 12c^2$

15. $5m^3 + 50m^2 + 125m$

16. $6y^3 + 2y^2 - 20y$

17. $6t^3 + 9t^2 - 15t$

18. $56x - 14x^2 - 21x^3$

19. $x^3 - 2x^2 + 3x - 6$

20. $5x^3 - 20x$

21. $t^3 + 3t^2 - 4t - 12$

22. $2x^3 + 3x^2 - 2x - 3$

23. $x^3 - 4x^2 + 3x - 12$

24. $2d^3 - 10d^2 + 3d - 15$

Solve the equation. Tell which solution method you used.

25. $21x^2 - 57x - 18 = 0$

26. $16x^2 + 25 = 0$

27. $2x^2 + 6x - 3 = 0$

28. $5x^2 + 4x + 3 = 0$

29. $3x^2 - 5x - 1 = 0$

30. $10x^2 - 38x + 36 = 0$

Vertical Motion **In Exercises 31–33, use the vertical motion models, where *h* is the initial height (in feet), *v* is the initial velocity (in feet per second), and *t* is the time (in seconds) the object spends in the air.**

Vertical motion model for Earth: $h = 16t^2 - vt$

Vertical motion model for the moon: $h = \frac{16}{6}t^2 - vt$

31. ***Earth*** You toss a baseball from a height of 64 feet with an initial upward velocity of 48 feet per second. How long will it take the baseball to reach the ground?

32. ***Moon*** On the moon, you toss a baseball from a height of 64 feet with an initial upward velocity of 48 feet per second. How long will it take the baseball to reach the surface of the moon?

33. Do objects fall faster on Earth or on the moon?

Lesson 10.8

NAME _____ DATE _____

Reteaching with Practice

For use with pages 625–632

GOAL Use the distributive property to factor a polynomial and solve polynomial equations by factoring

---

VOCABULARY

A polynomial is **prime** if it is not the product of polynomials having integer coefficients.

To **factor a polynomial completely,** write it as the product of monomial factors and prime factors with at least two terms.

EXAMPLE 1 *Finding the Greatest Common Factor*

Factor the greatest common factor out of $35x^3 + 45x^5$.

SOLUTION

First find the greatest common factor (GCF). It is the product of all the common factors.

$$35x^3 = 5 \cdot 7 \cdot x \cdot x \cdot x$$

$$45x^5 = 5 \cdot 9 \cdot x \cdot x \cdot x \cdot x \cdot x$$

$$\text{GCF} = 5 \cdot x \cdot x \cdot x = 5x^3$$

Use the distributive property to factor the greatest common factor out of the polynomial.

$$35x^3 + 45x^5 = 5x^3(7 + 9x^2)$$

Exercises for Example 1

Find the greatest common factor and factor it out of the expression.

1. $24y^3 + 32y$ **2.** $6n^8 - 18n^3$ **3.** $3a^2 + 30$

EXAMPLE 2 *Factoring Completely*

Factor $3x^4 + 30x^3 + 27x^2$ completely.

SOLUTION

$$3x^4 + 30x^3 + 27x^2 = 3x^2(x^2 + 10x + 9) \quad \text{Factor out GCF.}$$

$$= 3x^2(x + 9)(x + 1) \quad \text{Factor } x^2 + bx + c \text{ when } b \text{ and } c \text{ are positive.}$$

Exercises for Example 2

Factor the expression completely.

4. $2y^3 - 18y$ **5.** $7t^5 + 14t^4 + 7t^3$ **6.** $x^4 - 3x^3 + 2x^2$

Algebra 1
Chapter 10 Resource Book

NAME _____ DATE _____

Reteaching with Practice

For use with pages 625–632

EXAMPLE 3 *Factoring by Grouping*

Factor $x^4 - 3x^3 + 4x - 12$ completely.

SOLUTION

Sometimes you can factor polynomials that have four terms by grouping the polynomial into two groups of terms and factoring the greatest common factor out of each term.

$$
\begin{aligned}
x^4 - 3x^3 + 4x - 12 &= (x^4 - 3x^3) + (4x - 12) &&\text{Group terms.} \\
&= x^3(x - 3) + 4(x - 3) &&\text{Factor each group.} \\
&= (x - 3)(x^3 + 4) &&\text{Use distributive property.}
\end{aligned}
$$

Exercises for Example 3

Factor the expression completely.

7. $y^3 + 3y^2 - 2y - 6$ **8.** $x^3 + 2x^2 + 5x + 10$ **9.** $d^4 - d^3 + d - 1$

EXAMPLE 4 *Solving a Polynomial Equation*

Solve $7x^3 - 63x = 0$.

SOLUTION

$$
\begin{aligned}
7x^3 - 63x &= 0 &&\text{Write original equation.} \\
7x(x^2 - 9) &= 0 &&\text{Factor out GCF.} \\
7x(x + 3)(x - 3) &= 0 &&\text{Factor difference of two squares.}
\end{aligned}
$$

By setting each variable factor equal to zero, you find the solutions to be $0, -3$, and 3.

Exercises for Example 4

Solve the equation.

10. $y^2 - 4y - 5 = 0$ **11.** $3w^3 - 75w = 0$ **12.** $2x^3 + 12x^2 + 18x = 0$

Lesson 10.8

NAME _____ DATE _____

Quick Catch-Up for Absent Students

For use with pages 625–632

The items checked below were covered in class on (date missed) _____

Lesson 10.8: Factoring Using the Distributive Property

_____ **Goal 1:** Use the distributive property to factor a polynomial. (pp. 625–626)

Material Covered:

_____ Example 1: Finding the Greatest Common Factor

_____ Student Help: Skills Review

_____ Example 2: Recognizing Complete Factorization

_____ Example 3: Factoring Completely

_____ Example 4: Factoring Completely

_____ Example 5: Factoring by Grouping

_____ Example 6: Factoring by Grouping

Vocabulary:

prime factor, p. 625   factor a polynomial completely, p. 625

_____ **Goal 2:** Solve polynomial equations by factoring. (pp. 627–628)

Material Covered:

_____ Example 7: Solving a Polynomial Equation

_____ Example 8: Writing and Using a Polynomial Model

_____ Example 9: Solving Quadratic and Other Polynomial Equations

_____ Other (specify) _____

_____

Homework and Additional Learning Support

_____ Textbook (specify) _pp. 629–632_____

_____

_____ Internet: Extra Examples at www.mcdougallittell.com

_____ *Reteaching with Practice* worksheet (specify exercises)_____

_____ *Personal Student Tutor* for Lesson 10.8

NAME _____ DATE _____

Real-Life Application:
When Will I Ever Use This?

For use with pages 625–632

Playgrounds

A playground is an outdoor area set aside for play. Playgrounds were first started for children. But people of all ages enjoy the many different playground activities.

Small children can play in sand piles and on seesaws and swings. Older boys and girls may play or practice a variety of games and sports on the playground. Adults may participate in such games as tennis, badminton, and horseshoes.

Playgrounds often become a center of community activity. Parents come to watch competitive contests and other special events. Holiday events are often held on playgrounds. Many schools hold yearly play days on playgrounds.

In Exercises 1-4, use the following information.

You volunteer to help build a neighborhood playground. You are asked to dig a hole with a rectangular base. The height (or depth) of the hole is to be 2 feet less than the width. The length is to be 8 feet more than the width. The volume of dirt you will be digging up is 96 cubic feet.

1. Draw a diagram of the hole you are asked to dig. Label the length, width, and height.

2. Write a model for the volume of the hole. Solve the resulting equation.

3. What are the dimensions of the hole?

4. The hole you dug is to be filled with sand. Sand is sold for $40 per cubic yard. How much will the sand cost?

In Exercises 5 and 6, use the following information.

You purchase a plastic slide from a toy store. The slide is packaged in a box with a length that is 9 feet more than its width and a height that is 4 feet less than its width.

5. Find the dimensions of the box if its volume is 180 cubic feet.

6. The trailer you are pulling behind your vehicle has a length of 14 feet and a width of 7 feet. Will the box lie lengthwise in your trailer?

Challenge: Skills and Applications

For use with pages 625–632

In Exercises 1–4, factor the expression completely.

1. $4(x - 6)^3 + 17(x - 6)^2 - 15(x - 6)$

2. $x^8 - 256$

3. $(4x^2 - 20x + 25) - 81$

4. $8x^3 - 72x + 9x^2 - 81$

In Exercises 5–8, use factoring to solve the equation.

5. $(r + 4) = -2(r + 4)^2$

6. $4p^4 = 9p^2$

7. $8x^3 + 2x^2 + 44x^2 + 11x = 0$

8. $(r^2 - 7)^2 + 4 = 4(r^2 - 7)$

9. a. The polynomial $x^3 - 19x + 30$ can be factored into
   $(x + a)(x^2 + 2x - 15)$ for some value of a. What is that value?

b. Find the x-intercepts of the graph of the equation $x^3 - 19x + 30$.

In Exercises 10–12, use the following information.

A track runs along the outer edge of a circular park. The track is the same
width all the way around the park. The area inside the track is given by the
trinomial $\pi r^2 - 4\pi r + 4\pi$, where r is the radius of the entire park in meters.

10. Factor $\pi r^2 - 4\pi r + 4\pi$ completely.

11. What is the width of the track?

12. If the area of the park that is inside the track is 100π, what is the radius of
   the entire park?

Algebra 1
Chapter 10 Resource Book

Chapter Review Games and Activities

For use after Chapter 10

Solve the following problems, and find the answer at the right of the page. Place the letter of the answer on the line with the problem number to answer the riddle.

How does a pea farmer keep from getting sunburned at the beach?

1. $(-2x^3 + 2x^2 - x - 1) - (3x^3 + 5x^2 - x + 5) =$

2. $(x + 3)(-3x - 6) =$

3. $(4x - 8)(4x + 8) =$

4. $(5x + 2)^2 =$

5. Solve for x: $(x - 4)(x + 2) = 0$

6. Solve for x: $x^2 + 3x = 54$

7. Factor: $3x^2 - 19x - 40$

8. Factor: $9x^2 - 12x + 4$

(S) $x^3 + 7x^2 - 2x + 4$

(C) $x = -4$ and 2

(S) $(3x - 2)(3x - 2)$

(U) $3x^2 + 15x - 18$

(A) $x = 9$ and -6

(N) $16x^2 - 64$

(O) $25x^2 + 20x + 4$

(S) $16x^2 + 64$

(A) $(3x + 5)(x - 8)$

(R) $(3x - 2)(3x + 2)$

(E) $x = -9$ and 6

(E) $(9x - 2)(x - 2)$

(P) $x = -2$ and 4

(A) $-3x^2 - 15x - 18$

(N) $(-3x - 20)(x + 2)$

(M) $25x^2 + 4$

(C) $-5x^3 - 3x^2 - 6$

He uses ____ ____ ____ ____ ____ ____ ____ ____ !!
        1   2   3   4   5   6   7   8

NAME _____ DATE _____

Chapter Test A

For use after Chapter 10

Find the sum or difference.

1. $(x^2 + 2x + 1) + (4x^2 + 5x + 3)$

2. $(3x^2 + 5x + 4) - (x^2 + 2x + 1)$

3. $(3x^2 + 5x + 8) + (6x^2 + 3x + 2)$

4. $(8x^2 + 6x + 3) - (4x^2 + 3x + 2)$

5. The profits of a company are found by subtracting the company's costs from its revenue. If a company's cost can be modeled by $14x + 120,000$ and its revenue can be modeled by $40x - 0.0002x^2$, what is an expression for the profit?

Find the product.

6. $2x(3x - 5)$

7. $(x + 3)(x + 2)$

8. $(2x + 1)(x + 3)$

9. $(x + 1)(x^2 + x + 1)$

10. $(x + 2)(x - 2)$

11. $(x + 5)^2$

12. Find an expression for the area of the figure.

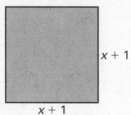

Use the zero-product property to solve the equation.

13. $(x + 4)(x - 2) = 0$

14. $(x + 3)^2 = 0$

Match the function with its graph.

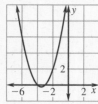 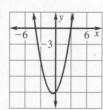 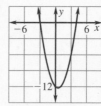

15. $y = (x - 3)(x + 4)$

16. $y = (x + 3)(x - 4)$

17. $y = (x + 3)(x + 4)$

1. _____
2. _____
3. _____
4. _____
5. _____
6. _____
7. _____
8. _____
9. _____
10. _____
11. _____
12. _____
13. _____
14. _____
15. _____
16. _____
17. _____

NAME _____ DATE _____

Chapter Test A

For use after Chapter 10

Factor the expression.

18. $x^2 + 3x + 2$

19. $x^2 - x - 6$

20. $x^2 - 7x + 10$

21. $x^2 + x - 12$

22. A rectangle has an area given by $A = x^2 + 5x + 6$. Find expressions for the possible length and width of the rectangle.

Factor the expression.

23. $2x^2 + 5x + 3$

24. $6x^2 + 14x + 4$

25. $x^2 - 9$

26. $x^2 - 8x + 16$

27. $x^2 + 6x + 9$

28. $2x^2 - 4x$

29. $x^3 + 2x^2 - 3x - 6$

30. $x^3 + 4x^2 + x + 4$

Write a quadratic equation that has the given solutions.

31. 3 and 2

32. 1 and 3

Solve the equation by a method of your choice.

33. $x^2 - 2x - 3 = 0$

34. $x^2 + 5x + 6 = 0$

35. $x^2 + 3x - 4 = 0$

36. $x^2 - 6x + 5 = 0$

37. $6x^2 + 5x + 1 = 0$

38. $4x^2 + 15x + 9 = 0$

39. $x^2 - 36 = 0$

40. $x^2 + 12x + 36 = 0$

18. _____

19. _____

20. _____

21. _____

22. _____

23. _____

24. _____

25. _____

26. _____

27. _____

28. _____

29. _____

30. _____

31. _____

32. _____

33. _____

34. _____

34. _____

35. _____

36. _____

37. _____

38. _____

39. _____

40. _____

Review and Assess

NAME _____ DATE _____

Chapter Test B

For use after Chapter 10

Find the sum or difference.

1. $(2x^2 + 3x + 5) + (-x^2 + 4x - 7)$

2. $(5x^2 + 8x + 5) - (2x^2 - 3x + 2)$

3. $(9x^2 + 8x + 4) + (8x^2 - 4x - 8)$

4. $(14x^2 - 5x + 2) - (3x^2 - 8x - 4)$

5. The profits of a company are found by subtracting the company's costs from its revenue. If a company's cost can be modeled by $18x + 110,000$ and its revenue can be modeled by $60x - 0.0003x^2$, what is an expression for the profit?

Find the product.

6. $(4x + 2)(6x^2)$

7. $(x - 4)(x - 5)$

8. $(4x - 5)(x + 2)$

9. $(x - 3)(x^2 + x + 1)$

10. $(x + 5)(x - 5)$

11. $(x - 8)^2$

12. Find an expression for the area of the figure.

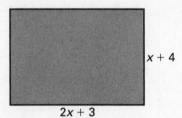

$x + 4$

$2x + 3$

Use the zero-product property to solve the equation.

13. $(x + 5)(x - 1) = 0$

14. $(2x + 1)^2 = 0$

Match the function with its graph.

A 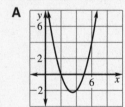   B 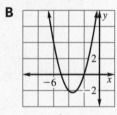   C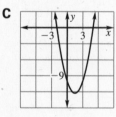

15. $y = (x + 5)(x + 2)$

16. $y = (x - 5)(x + 2)$

17. $y = (x - 5)(x - 2)$

1. _____

2. _____

3. _____

4. _____

5. _____

6. _____

7. _____

8. _____

9. _____

10. _____

11. _____

12. _____

13. _____

14. _____

15. _____

16. _____

17. _____

Review and Assess

Chapter Test B

For use after Chapter 10

Factor the expression.

18. $x^2 + 11x + 30$ **19.** $x^2 - 3x - 40$

20. $x^2 - 9x + 14$ **21.** $x^2 + 2x - 35$

22. A rectangle has an area given by $A = x^2 + 7x + 12$. Find expressions for the possible length and width of the rectangle.

Factor the expression.

23. $5x^2 + 16x + 3$ **24.** $12x^2 + 14x + 4$

25. $x^2 - 16$ **26.** $x^2 - 12x + 36$

27. $x^2 + 10x + 25$ **28.** $3x^3 + 9x^2 + 2x$

29. $x^3 - 3x^2 + 2x - 6$ **30.** $x^3 - x^2 + 5x - 5$

Write a quadratic equation that has the given solutions.

31. 5 and 3   **32.** 7 and 11

Solve the equation by a method of your choice.

33. $x^2 - 2x - 8 = 0$ **34.** $x^2 + 9x + 20 = 0$

35. $x^2 + 4x - 12 = 0$ **36.** $x^2 - 11x + 30 = 0$

37. $4x^2 + 5x + 1 = 0$ **38.** $4x^2 + 10x + 6 = 0$

39. $x^2 - 49 = 0$ **40.** $x^2 - 8x + 16 = 0$

18. _____

19. _____

20. _____

21. _____

22. _____

23. _____

24. _____

25. _____

26. _____

27. _____

28. _____

29. _____

30. _____

31. _____

32. _____

33. _____

34. _____

35. _____

36. _____

37. _____

38. _____

39. _____

40. _____

Review and Assess

Chapter Test C

For use after Chapter 10

Find the sum or difference.

1. $(-4x^2 + 9x - 12) + (3x^2 - 4x - 8)$

2. $(9x^2 - 3x + 4) - (-x^2 + 3x - 5)$

3. $(5x^2 - 2x - 3) + (7x^2 - 6x + 4)$

4. $(13x^2 - 12x - 5) - (-4x^2 + 11x - 10)$

5. The profits of a company are found by subtracting the company's costs from its revenue. If a company's cost can be modeled by $10x + 130{,}000$ and its revenue can be modeled by $50x - 0.0004x^2$, what is an expression for the profit?

Find the product.

6. $(9x^2 + 3x - 5)(-4x)$

7. $(x - 8)(x + 4)$

8. $(3x - 4)(9x + 5)$

9. $(2x + 4)(x^2 - x + 1)$

10. $(2x + 3)(2x - 3)$

11. $(5x + 2)^2$

12. Find an expression for the area of the figure.

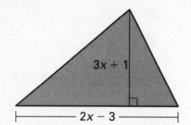

$3x + 1$

$2x - 3$

Use the zero-product property to solve the equation.

13. $(2x + 8)(3x - 6) = 0$

14. $(4x + 12)(2x - 2)(3x - 9) = 0$

Match the function with its graph.

A
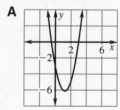

B
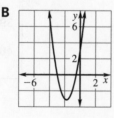

C
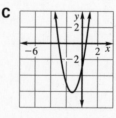

15. $y = (2x + 1)(x + 3)$

16. $y = (2x - 1)(x + 3)$

17. $y = (2x + 1)(x - 3)$

1. _____
2. _____
3. _____
4. _____
5. _____
6. _____
7. _____
8. _____
9. _____
10. _____
11. _____
12. _____
13. _____
14. _____
15. _____
16. _____
17. _____

Review and Assess

Chapter Test C

For use after Chapter 10

Factor the expression.

18. $x^2 + 15x + 56$

19. $x^2 - 2x - 63$

20. $x^2 - 8x + 12$

21. $x^2 + 3x - 28$

22. A rectangle has an area given by $A = x^2 - 3x - 10$. Find expressions for the possible length and width of the rectangle.

Factor the expression.

23. $3x^2 + 5x - 2$

24. $10x^2 + 4x - 6$

25. $4x^2 - 25$

26. $4x^2 - 20x + 25$

27. $9x^2 + 42x + 49$

28. $6x^3 - 15x^2 - 9x$

29. $x^3 + 3x^2 - 4x - 12$

30. $x^3 + 5x^2 - 16x - 80$

Write a quadratic equation that has the given solutions.

31. -2 and 5

32. -3 and -5

Solve the equation by a method of your choice.

33. $x^2 - 2x - 15 = 0$

34. $x^2 + 12x = -36$

35. $x^2 + 3x - 40 = 0$

36. $x^2 - 9x + 18 = 0$

37. $6x^2 - 4x - 2 = 0$

38. $9x^2 - 5x - 4 = 0$

39. $16x^2 - 25 = 0$

40. $25x^2 + 50x + 25 = 0$

18. _____

19. _____

20. _____

21. _____

22. _____

23. _____

24. _____

25. _____

26. _____

27. _____

28. _____

29. _____

30. _____

31. _____

32. _____

33. _____

34. _____

35. _____

36. _____

37. _____

38. _____

39. _____

40. _____

Review and Assess

NAME _____ DATE _____

SAT/ACT Chapter Test

For use after Chapter 10

1. Which of the following is equal to $(7x^3 - 3x^2 + 5x - 5) + (5x^2 - 8x - 3)$?

   Ⓐ $7x^3 + 2x^2 - 3x - 8$

   Ⓑ $7x^3 - 11x^2 + 5x - 8$

   Ⓒ $10x^3 - 3x^2 - 3x - 8$

   Ⓓ $12x^3 + 2x^2 - 3x - 8$

2. Which trinomial represents the area of the triangle?

   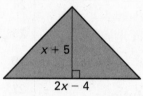

   $x + 5$

   $2x - 4$

   Ⓐ $x^2 + \frac{1}{2}x - 10$

   Ⓑ $x^2 + 3x - 10$

   Ⓒ $2x^2 + 6x - 20$

   Ⓓ $2x^2 + x - 20$

3. Which of the following is equal to $(3x + 7)^2$?

   Ⓐ $9x^2 + 42x + 49$ Ⓑ $9x^2 + 21x + 49$

   Ⓒ $9x^2 + 49$ Ⓓ $9x^2 + 49x + 21$

4. What are the x-intercepts of the graph of $y = (2x + 3)(x - 5)$?

   Ⓐ $\frac{3}{2}$ and -5 Ⓑ $-\frac{3}{2}$ and -5

   Ⓒ $\frac{3}{2}$ and 5 Ⓓ $-\frac{3}{2}$ and 5

5. A triangle's base is 14 inches less than 2 times its height. If h represents the height in inches, and the total area of the triangle is 54 square inches, which of the following equations can be used to determine the height?

   Ⓐ $7h - h^2 = 54$ Ⓑ $2h^2 - 14h = 54$

   Ⓒ $h^2 - 7h = 54$ Ⓓ $14h - 2h^2 = 54$

6. Which of the following equations *cannot* be solved by factoring with integer coefficients?

   Ⓐ $3x^2 + 11x - 15 = 0$

   Ⓑ $6x^2 + 2x - 20 = 0$

   Ⓒ $12x^2 + 13x - 14 = 0$

   Ⓓ $18x^2 + 3x - 10 = 0$

In Questions 7 and 8, choose the statement below that is true about the given numbers.

   A The number in column A is greater.

   B The number in column B is greater.

   C The two numbers are equal.

   D The relationship cannot be determined from the given information.

7.

Column A	Column B
$(2x + 3y)^2$ when $x = 5$ and $y = -4$	$(2x)^2 + (3y)^2$ when $x = 5$ and $y = -4$

   Ⓐ   Ⓑ   Ⓒ   Ⓓ

8.

Column A	Column B
$(5x - 2y)^2$ when $x = -6$ and $y = 2$	$(5x)^2 - (2y)^2$ when $x = -6$ and $y = 2$

   Ⓐ   Ⓑ   Ⓒ   Ⓓ

9. Which of the following is equal to the expression $3x^3 + 15x^2 + 4x + 20$?

   Ⓐ $(3x + 4)(x + 5)$

   Ⓑ $(3x + 4)(x^2 + 5)$

   Ⓒ $(3x^2 + 4)(x + 5)$

   Ⓓ $(3x^2 + 5)(x + 4)$

NAME _____ DATE _____

Alternative Assessment and Math Journal

For use after Chapter 10

JOURNAL

1. You have learned about solving quadratic equations by various methods. Remember there are three methods that you have learned to solve quadratic equations. (a) Consider the following quadratic equations. Determine the number of solutions of each quadratic equation.

   i. $x^2 - 4x + 3 = 0$ *ii.* $2x^2 - 4x - 3 = 0$ *iii.* $2x^2 + x + 3 = 0$

   (b) For each equation answer the following: Can the equation be solved by all three methods? Why or why not? Of the three methods for solving quadratic equations, which would you use to solve the equation? Explain your reasoning. (c) In general, state the advantages and disadvantages of each method for solving a quadratic equation. You may either write a paragraph to explain your reasons, or make a chart containing the advantages and disadvantages for each method.

MULTI-STEP PROBLEM

2. Consider the following figures to answer the questions that follow. Solve all quadratic equations by **factoring.**

Figure 1

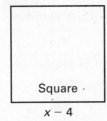

Square ·

$x - 4$

Figure 2

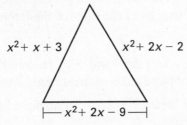

$x^2 + x + 3$ $x^2 + 2x - 2$

$\longmapsto x^2 + 2x - 9 \longmapsto$

Figure 3

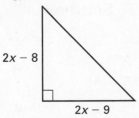

$2x - 8$

$2x - 9$

a. Write an expression for the perimeter of Figure 1 and for the perimeter of Figure 2.

b. Find the perimeter of Figure 1 if $x = 5$ centimeters and if $x = 3$ centimeters, if possible. If not possible, explain why not.

c. Find x if the perimeter of Figure 1 is 32 centimeters.

d. For Figure 2, find x if the perimeter is 4 centimeters. Does each answer make sense? Why or why not?

e. Write and simplify an expression for the area of Figure 1 and for the area of Figure 3.

f. Find the area of Figure 3 if $x = 10$ centimeters.

g. Determine x in Figure 1 if the area is 4 centimeters.

h. Determine the value(s) of x when the areas of Figures 1 and 3 are equal.

3. **Writing**   Clearly explain why the equations in parts (d), (g), and (h) of Exercise 2 had to be set equal to zero before you could solve. Use complete sentences. You may use examples to support your explanation.

Review and Assess

Alternative Assessment Rubric

For use after Chapter 10

JOURNAL
SOLUTION

1 a–c. Complete answers should address these points.

a. • Equations *i* and *ii* each have 2 solutions and *iii* has no solution.

b. • Explain that problem *i* can be solved by all three methods, *ii* and *iii* can only be solved by quadratic formula and graphing. Explain reasons why.

• Explain that *i* can be easily factored since its discriminant is a perfect square; *ii* cannot be factored but quadratic formula could be used to solve.

• Explain that in *ii* graphing would be harder to obtain an exact answer.

• Explain that part *iii* is most easily solved by graphing because one can quickly tell that there are no *x*-intercepts.

c. • Explain advantages and disadvantages including exact versus approximate answers for graphing, factorability and how easy or difficult a problem may be to factor.

MULTI-STEP
PROBLEM
SOLUTION

2 a. Figure 1 perimeter: $4x - 16$ units; Figure 2 perimeter: $3x^2 + 5x - 8$ units

b. 4 cm; -4 cm which is not possible. *Sample Answer:* One must make sure that the lengths of the sides of the figure are positive numbers.

c. 12

d. Solutions for *x* are $\frac{4}{3}$ and -3. However, neither is true because the length of one or more sides of the triangle would be negative.

e. Figure 1 area: $x^2 - 8x + 16$ square units; Figure 3 area: $2x^2 - 17x + 36$ square units

f. 66 cm^2

g. 6 (*x* cannot be 2, otherwise sides would be negative.)

h. 5 (*x* cannot be 4, otherwise sides of square would be 0 units long.)

3. *Sample Answer:* To use zero-product property, equation must equal zero.

MULTI-STEP
PROBLEM
RUBRIC

4 Students complete all parts of the questions accurately. Explanations are logical and clear. Statements are included as to why answers are not true based on the geometric shape. Students solve the problems by the factoring method. Proper units are included in the solutions.

3 Students complete the questions and explanations. Solutions may contain minor mathematical errors or misunderstandings. Statements are included as to why answers are not true based on the geometric shape. Students solve the problems by the factoring method. Most answers include proper units.

2 Students complete questions and explanations. Several mathematical errors may occur. No explanation is given why answers are not true based on the geometric shape. Students do not always use the factoring method to solve. Units are not given.

1 Students' work is very incomplete. Solutions and reasoning are incorrect. No explanation is given as to why answers are not true based on the geometric shape. Students do not use the factoring method to solve. Units are not given.

Algebra 1
Chapter 10 Resource Book

NAME _____ DATE _____

Project: Is it Realistic?

For use with Chapter 10

OBJECTIVE Determine how much a doll or action figure would weigh if it were a real person.

MATERIALS a doll or action figure, a pitcher large enough to contain the doll or action figure, a large pot, measuring cup, measuring spoons, tape measure, tape, paper, pencil, and calculator

INVESTIGATION Dolls and action figures are scale models of human beings, but are their dimensions realistic? You can determine how accurate a figure's proportions are by measuring its volume.

Measure the volume of the doll or action figure by measuring the amount of water it displaces. Place the pitcher inside the pot. Fill the pitcher to the very top with water, being careful not to spill any water into the pot. Tape over any places where water might leak in if your figure is hollow. Completely submerge the figure in the water without letting your fingers go beneath the surface. Remove the figure and the pitcher from the pot. Measure the volume of water that spilled over into the pot. This is the volume of the doll or action figure.

1. Convert the volume of the figure to liters using the conversion chart.

U.S.	*Metric Units*
1 cup	0.24 liters
1 tablespoon	0.015 liters
1 teaspoon	0.005 liters

2. The volume of a real person V_p is related to the volume of a scale model V_m in liters by $V_p = V_m s^3$, where s is a scale factor given by

$$s = \frac{\text{Person's height (in.)}}{\text{Model's height (in.)}}.$$ Measure the height of the doll or action figure.

Then decide how tall the doll or action figure would be if it were a real person and find s. What would be the person's volume?

3. A person's weight is given by

weight = volume × weight density

The average weight density for females is 2.31 pounds per liter. The average weight density for males is 2.34 pounds per liter. Find how much your figure would weigh if it were a real person. Is it a realistic weight?

PRESENT YOUR RESULTS Write a report about your experiment. Describe the procedures you used, the data you collected, your calculations, and some possible sources of error. Analyze your results. Is the weight you calculated realistic? Why or why not? If not, what could the designers of the figure do to make it more realistic?

Review and Assess

Project: Teacher's Notes

For use with Chapter 10

GOALS • Evaluate expressions containing exponents and use exponents in real-life problems.

• Use polynomials to model real-life situations.

• Analyze the relationship between a scale model and a real object.

MANAGING THE PROJECT You may wish to have students work in groups of 2 to 4. Encourage groups to collect the volume data carefully and to repeat the procedures at least twice. Important points to address are filling the pitcher without spilling any water in the pot and measuring the displaced water carefully.

RUBRIC The following rubric can be used to assess student work.

4 The student measures and performs calculations carefully and accurately. The report describes procedures thoroughly, presents data collected, shows step-by-step calculations, and describes possible sources of error. The discussion of realism of the figure shows insight and understanding of the issues. For example, there may be a discussion of the materials that the doll or action figure is made of as compared with the physiology of a human being.

3 The student measures and performs calculations with few, if any, errors. The report describes procedures, presents the data collected, shows step-by-step calculations, and describes possible sources of error. However, the presentation is not as thorough as possible or does not show complete understanding of the issues.

2 The student measures and performs calculations. However, work may be incomplete or reflect misunderstanding. For example, procedures may not be described thoroughly or appropriately, data may not be presented, or calculations may have errors. The report may indicate a limited grasp of the issue of realism of the figure.

1 The description of procedures, data collected, or calculations are missing or incorrect. The report does not give an appropriate discussion of the realism of the figure.

NAME _____ DATE _____

Cumulative Review

For use after Chapters 1–10

Write the verbal phrase as an algebraic expression. Use *x* for the variable in your expression. (1.5)

1. Ten more than a number

2. Product of one third and a number

3. Difference of eight and a number

4. Five cubed divided by a number

Evaluate the expression for the given value. (2.5)

5. $2y^3 - 10$ when $y = -2$

6. $-(-b^4)(b)(b^2)$ when $b = -3$

7. $\frac{7}{10}(-w)(10w^2)$ when $w = 4$

8. $|5 - t^3| - |t^3|$ when $t = 6$

In Exercises 9–11, decide whether the relation is a function. If it is, give the domain and the range. (4.8)

9. Input   Output

1 ⟶ 8
2 ⟶ 9
3 ⟶ 13
4 ⟶ 20

10. Input   Output

11. Input   Output

10 ⟶ 2
12 ⟶ 3
18 ⟶ 5
22

Graph the line that passes through the points. Write its equation in slope-intercept form. (5.3)

12. $(3, 8), (9, 0)$

13. $(7, 9), (-6, 9)$

14. $(-3, -2), (-8, 1)$

15. $(1, -2), (5, -8)$

16. $(12, 0), (0, 8)$

17. $(-6, -3), (-9, 2)$

Make a box-and-whisper plot of the data. (6.7)

18. 3, 5, 6, 6, 8, 9

19. 1, 1, 2, 3, 5, 10

20. 4, 7, 2, 3, 5, 6, 20

Use the substitution method or linear combinations to solve the linear system. (7.5)

21. $7 = 5x - y$

$x = -4y + 16$

22. $6x + y - 8 = 0$

$-4x = 9 - 2y$

23. $16x - 20y = -400$

$2x = 8 - 3y$

Simplify the expression. The expression should have no negative exponents. (8.3)

24. $\left(\dfrac{2}{x}\right)^{-2}$

25. $\left(\dfrac{y^3}{x^{-5}}\right)^{-9}$

26. $\left(\dfrac{6x^3y^{-7}}{9xy}\right)^{-1}$

27. $\dfrac{(r^{-3})^5}{(r^{-4})^8}$

28. $\left(\dfrac{m^{-3}}{m^7m^{-4}}\right)^{-3}$

29. $\left(\dfrac{15x^3}{6x^{-5}}\right)^{-1} \cdot x^{-10}$

Review and Assess

Use the quadratic formula to solve. (9.5)

30. $y^2 - 12y + 1 = 0$

31. $x^2 - 15x + 5 = 0$

32. $7x^2 + x + 1 = 0$

33. $0.2x^2 - 0.3x - 0.5 = 0$

Tell if the equation has *two solutions, one solution, or no solution*. (9.6)

34. $5x^2 - 9x + 16 = 0$

35. $x^2 = 2x - 1$

36. $3x^2 - x - 1 = 0$

Use a vertical format or a horizontal format to find the sum or difference. (10.1)

37. $(8x^4 + 11) + (12x^4 + x - 10)$

38. $(-2t^3 + 4t^2 - 4) - (4t^3 + 5t^2 + t + 6)$

39. $\left(x^4 - \frac{5}{3}x^2\right) + (x^2 - 2) - \left(\frac{2}{3}x^2 - 1\right)$

40. $(4.8m^4 + 2.3m) - (4.23m^4 + 23.2m)$

Use the FOIL pattern to find the product. (10.2)

41. $(r - 2)(3r + 4)$

42. $(2y - 7)(3y + 1)$

43. $(6t - 9)(t + 2)$

44. $(7 - 3x)(4 - 2x)$

45. $(q - 4)(2.2q + 0.5)$

46. $\left(s - \frac{1}{2}\right)\left(s + \frac{1}{4}\right)$

Write the square of the binomial as a trinomial. (10.3)

47. $(x + 6)^2$

48. $(2y - 1)^2$

49. $(4s + 2)^2$

50. $(x - 2.3)^2$

51. $\left(\frac{1}{4}x - 1\right)^2$

52. $(5w - 1.2)^2$

Solve the equation. (10.4)

53. $(3x - 2)(x + 1)(2x - 5) = 0$

54. $(6b - 3)(4b - 3)(8b - 1) = 0$

55. $(9x - 4.1)^2(7x - 5.4)$

56. $\left(9n - \frac{8}{9}\right)(n - 2)\left(\frac{1}{6}n + 1\right)$

Factor the trinomial. (10.5–10.7)

57. $4n^2 + 4n + 1$

58. $16x^2 - 49$

59. $2x^2 + 3x - 9$

60. $6x^2 + 8x + 2$

61. $3y^2 - 8y - 3$

62. $\frac{1}{4}x^2 + \frac{1}{2}x - 6$

Factor the expression completely. (10.8)

63. $24x^2 - 30$

64. $7a^2 - 28$

65. $18u^2 + 9u$

66. $-x^4 + 5x^2$

67. $t^3 - t^2 - 81t + 81$

68. $c^4 + 2c^3 + c + 2$

ANSWERS

Chapter Support

Parent Guide
Chapter 10

10.1: $6x^3 + 3x - 2; 2x^3 - 7x + 16$
10.2: $R = 0.6x^2 + 19.5x + 150$
10.3: $38 \cdot 42 = (40-2)(40+2) = 1600 - 4 = 1596$
10.4: 24 ft; 18 ft **10.5:** $7, -6$ **10.6:** 1 sec
10.7: $(x - 12)$ft by $(x - 12)$ft
10.8: $4y(x - 9)(x - 2)$

Prerequisite Skills Review

1. $10 + 20x$ **2.** $-\frac{1}{4}t + 4$ **3.** $-2y + 6$

4. $9s - 20$ **5.** $4.2x + 23.04$ **6.** $x^2 - 3x$

7. $x^6 y^{17}$ **8.** $-12a^7 b^{13}$ **9.** $192s^{15}t^{31}$

10. $\frac{1}{32}x^{28}y^{32}$ **11.** no solution **12.** one solution

13. two solutions **14.** no solution

Strategies for Reading Mathematics

1. Answers may vary. *Sample answers:* a unit of temperature, the extent or measure of a condition or situation, one of a series of steps, a unit of measure of an angle; you can generally tell by the context which meaning is being used.

2. If you don't know that the *nonnegative integers* are integers 0, 1, 2, 3, and so on, you might mistakenly think that $x^{-4} + 2x$ is a polynomial. Using the word *nonnegative* rather that *positive* allows the inclusion of zero as an exponent in a polynomial.

3. Each of the terms should be familiar, having been defined previously in the text. Each can be looked up in the Glossary or Index, if necessary.

Lesson 10.1

Warm-up Exercises

1. $4; -2$ **2.** $\frac{2}{3}; \frac{4}{3}$ **3.** $2; -8$ **4.** $-5x - 2$

5. $-2x - 4$

Daily Homework Quiz

1. 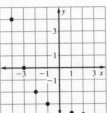 **2.**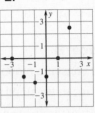

exponential   quadratic

3. $P = 24.5 \cdot 1.1^t$, where t is the number of years after 1995

Lesson Opener
Allow 10 minutes.

1. C **2.** Add like terms; $16 + 4x$ **3.** A

4. Clear parentheses and subtract like terms; $25 + 5x$

Practice A

1. $4x^2 + 3x - 5$ **2.** $5x^2 - 3x + 4$

3. $-7x^3 + x + 2$ **4.** $4x^2 + 2x + 8$

5. $4x^3 + 5x^2 - 2x$ **6.** $7x^4 - 5x^3 - 4x + 1$

7. $2x^2 + 3x - 7$ **8.** $7x - 2$

9. $x^3 + 2x^2 - x - 2$

10. 14; constant, monomial

11. 2; linear, binomial

12. -3; quadratic, trinomial

13. 1; cubic, binomial

14. -1; quartic, binomial

15. -1; quartic, polynomial

16. 2; quadratic, trinomial

17. 1; quadratic, binomial

18. -1; cubic, polynomial **19.** $5x^2 + 3x + 4$

20. $3x^2 + x + 3$ **21.** $5n^2 + n + 1$

22. $n^2 + 3n + 4$ **23.** $2a^3 - 6a^2 + a + 4$

24. $-4n^2 - 7n + 2$ **25.** $x^2 + 3x - 2$

26. $5m^2 - 2m + 1$ **27.** $x^2 + 4x + 6$

28. $2x^2 + 5x$

29. $7x^3 + x^2 + 5x + 6$ **30.** $6n^2 - 5n + 1$

31. $2x^3 + 7x^2 + 4x + 2$ **32.** $3x^2 - 10x + 10$

33. $20x - 1$ **34.** x^3 **35.** $5x^2 - 7x - 1$

36. $4x^2 - 6x + 5$ **37.** $3x^2 - 12\pi x + 8\pi$

38. $\frac{9}{4}x^2 + 6x - 30$

Practice B

1. 12; constant, monomial

2. 5; linear, binomial

3. -4; quadratic, trinomial

4. 5; cubic, binomial

5. -3; quartic, binomial

6. 2; quartic, polynomial

7. -3; cubic, binomial

8. 5; quadratic, trinomial

9. -1; quintic, polynomial

10. $2x^2 - 3x - 8$ **11.** $-x^3 - 2x^2 - 3x + 2$

12. $-4x^3 - 14x^2 + 3x + 13$ **13.** $4x^3 + 3x - 1$

14. $2n^3 + 4n^2 - 8$ **15.** $x^2 - 4x + 10$

16. $-3x^2 + 6$ **17.** $2x^2 + 5x + 2$

18. $6t^2 - 13t + 9$ **19.** $9x^3 + x^2 + 6x - 10$

20. $x^3 + 6x^2 - 6x + 11$ **21.** $5x^2 - x + 19$

22. $5x^3 - 7x^2 - 1$ **23.** $2x^3 + x^2 - 3x + 4$

24. $m^5 + 5m^3 + m - 4$ **25.** $-3x^2 - 11$

26. $7x^2 - 3x - 4$ **27.** $x^2 - 14x + 11$

28. $P = \frac{4}{15}t^2 + \frac{7}{3}t + 100$

29. $B = 0.014t^2 + 0.15t + 10$

Practice C

1. -4; constant, monomial

2. 3; linear, binomial

3. -1; quadratic, trinomial

4. 7; cubic, binomial **5.** -2; cubic, binomial

6. 2; quintic, polynomial

7. 3; quartic, binomial

8. -1; quadratic, trinomial

9. -1; quartic, polynomial

10. $8a^2$ **11.** $-4x^3 - x^2 - x + 4$

12. $-2x^3 + 2x^2 + 2x - 2$ **13.** $2x^4 + 4x - 1$

14. $10m^2 + 3m + 7$ **15.** $4n^5 + n^3 - 7$

16. $13x^2 + 2x - 1$ **17.** $4n^2 + 8n + 14$

18. $-n^2 - n + 2$ **19.** $9t^3 - 3t^2 + t - 3$

20. $x^3 + 11x^2 - 5x + 11$ **21.** $-3x + 5$

22. $3n^3 - 2n^2 - n - 7$ **23.** $x^3 + x^2 + 2x$

24. $-m^5 + 8m^3 + 3m - 1$ **25.** $x^2 - 1$

26. $x - 24$ **27.** $x^2 - 2x - 6$

28. $3x^3 - 2x^2 - 2x - 1$ **29.** $-3n^2 + n - 4$

30. $4t^2 + 20$ **31.** $-5u - 36$

32. $N = -0.47t^2 + 28.7t + 14.4$

33. $P = \frac{1}{6}t^2 + \frac{5}{6}t + 14$

Reteaching with Practice

1. $-3x^2 - x + 2$ **2.** $-4x^2 + 9x - 8$

3. $8x^2 - 2x - 1$ **4.** $-7x^2 + 3x - 3$

5. $N = 189t^2 - 983t - 307$

Interdisciplinary Application

1. $[(12x + 2 + x + x)](6x + x + x)]$
$- [(12x + 2)(6x)]$

2. $40x^2 + 4x$ **3.** 372 square inches

4. $6x(12x + 2); 72x^2 + 12x$

5. length = 50 inches; height = 24 inches

6. length = 44 inches; height 24 inches

7. $10,560

Challenge: Skills and Applications

1. -2 **2.** $\frac{1}{8}$ **3.** $-\frac{3}{5}$ **4.** $a = 3, b = -1$

5. $a = -2, b = -5$ **6.** -4 **7.** 9 **8.** 7

9. $-7x^3 + 7x^2 + 10x - 14$

10. $9x^3 - 17x^2 + 7x + 8$ **11.** $36x^2 + 12x$

Lesson 10.2

Warm-up Exercises

1. $2x + 8$ **2.** $15x - 6$ **3.** $21x + 74$

4. $-12x^2$ **5.** $-x$

Daily Homework Quiz

1. 5; quadratic, trinomial **2.** -3; linear, monomial **3.** $3x^2 + 3x + 2$ **4.** $x^2 + 3x - 11$

5. $b^4 + b^3$ **6.** $2y^2 + 8y + 10$

Lesson 10.2 *continued*

Lesson Opener

Allow 10 minutes.

1. D; x^2 2. A; $x^2 + x$ 3. F; $2x^2$
4. E; $2x^2 + 2x$ 5. B; $x^2 + 3x + 2$
6. C; $x^2 + 4x + 3$

Graphing Calculator Activity

1. correct 2. incorrect 3. incorrect
4. correct 5. $x^2 + 9x + 8$ 6. $x^2 - x - 90$
7. $3x^2 - 19x + 20$ 8. $9x^2 + 16x - 4$
9. $35x^2 + 43x + 12$ 10. $32x^2 + 20x - 18$

11. Answers will vary. *Sample answer:*
Distribute each term of the first binomial through each term of the second binomial. Then combine like terms to simplify

Practice A

1. $(x + 1)(2x + 2) = 2x^2 + 4x + 2$
2. $(x + 2)(2x + 3) = 2x^2 + 7x + 6$
3. $6x + 2$ 4. $-12x + 20$ 5. $10x^2 - 2x$
6. $24n - 30n^2$ 7. $3x^3 - 7x^2$
8. $24m^4 - 12m^3 + 3m^2$ 9. $-5t^3 - 10t^2 + 20t$
10. $6x^4 - 12x^3 - 21x^2$
11. $-10a^4 - 6a^3 + 14a^2$ 12. $t^2 + 6t + 9$
13. $n^2 + 6n + 5$ 14. $2x^2 - 3x - 20$
15. $8a^2 - 2a - 15$ 16. $3x^3 + 11x^2 + 7x + 3$
17. $8x^3 + 14x^2 - 11x + 10$ 18. $w^2 + 7w + 10$
19. $3z^2 + 7z + 2$ 20. $x^2 - 5x + 6$
21. $4x^2 + 27x + 35$ 22. $2x^2 + 14x - 16$
23. $20n^2 + 2n - 6$ 24. $6b^2 - 13b + 6$
25. $15x^2 + 2x - 8$ 26. $30n^2 - 5n - 10$
27. $3x^2 - 12x - 63$ 28. $16t^2 + 24t + 9$
29. $x^3 + x^2 - 5x - 2$ 30. $6x^2 + 10x - 4$
31. $x^2 + \frac{11}{2}x - \frac{21}{2}$

Practice B

1. $10x + 15$ 2. $-35x + 21$ 3. $18x^2 - 24x$
4. $12x^3 - 30x^2 + 6x$ 5. $-4x^3 + 7x^2$
6. $-18m^5 + 6m^3$ 7. $2x^5 - 2x^4 + 16x^3 - 10x^2$
8. $-15x^3 + 18x^2 + 24x$
9. $-15a^4 + 35a^3 - 45a^2$ 10. $2x^2 + x - 3$
11. $t^2 + 4t + 4$ 12. $n^2 + 6n + 8$

13. $9a^2 - 49$ 14. $2x^3 + 3x^2 - 17x + 12$
15. $6x^3 - 13x^2 + 7x + 4$ 16. $m^2 + 8m + 7$
17. $2t^2 + 7t + 3$ 18. $x^2 - 6x + 8$
19. $3x^2 + 14x + 16$ 20. $5x^2 + 32x - 21$
21. $30n^2 - 13n - 3$ 22. $3x^2 + 17x + 10$
23. $x^2 - x - 30$ 24. $x^2 - 12x + 32$
25. $x^2 - 3x - 28$ 26. $8x^2 + 5x - 3$
27. $5x^2 - 32x + 12$ 28. $x^3 + 4x^2 - 3x - 12$
29. $x^3 + 9x^2 + 20x$ 30. $2x^2 + \frac{29}{2}x + 15$
31. $\frac{1}{6}x^2 + x - 12$ 32. $2x^2 + \frac{2}{3}x - \frac{1}{6}$
33. $6x^2 + 19x + 10$ 34. $12x^2 - 20x + 7$
35. $25x^3 + 10x^2 - 10x - 4$
36. $12x^2 - 23x - 9$ 37. $6x^3 + 2x^2 + 12x + 4$
38. $12x^2 + 8x - \frac{5}{3}$ 39. $32x^3 - 8x^2 + 12x - 3$
40. $6x^2 + 9x - 2$; 2578 ft^2
41. $18x^2 + 30x + 12$; 144 in.^{3}

Practice C

1. $21x + 14$ 2. $-36x + 54$ 3. $15x^2 - 21x$
4. $-10x^3 - 30x^2 + 15x$ 5. $-15x^3 + 12x^2$
6. $-32x^5 + 4x^3$ 7. $5x^5 - 15x^4 + 10x^3 - 5x^2$
8. $6x^3 + 9x^2 - 12x$ 9. $-x^4 - 2x^3 + 3x^2$
10. $5x^2 - 18x - 8$ 11. $m^2 - 25$
12. $t^2 + 12t + 32$ 13. $8n^2 - 6n - 35$
14. $3x^3 + 11x^2 - 19x + 5$
15. $8x^3 - 10x^2 + 13x + 4$
16. $a^2 + 8a + 15$ 17. $2t^2 + 17t + 35$
18. $x^2 - 10x + 24$ 19. $15x^2 - 7x - 2$
20. $2x^2 + x - 6$ 21. $12n^2 - 4n - 1$
22. $x^2 + 7x - 8$ 23. $x^2 - 36$
24. $x^2 - 12x + 32$ 25. $15x^2 + 29x - 14$
26. $4x^2 + 27x - 40$ 27. $7x^2 - 26x + 15$
28. $x^3 - 5x^2 + 4x - 20$ 29. $3x^3 + 19x^2 + 20x$
30. $2x^2 - \frac{47}{3}x - 21$ 31. $\frac{1}{4}x^2 + 2x - 32$
32. $x^2 + \frac{1}{6}x - \frac{1}{18}$ 33. $14x^2 - 53x + 14$
34. $7.75x^2 + 8.1x + 2$ 35. $x^2 + \frac{1}{2}x - \frac{3}{16}$
36. $3x^2 - 23x - 8$ 37. $10x^3 - 15x^2 + 6x - 9$
38. $35x^3 - \frac{7}{2}x^2 + 10x - 1$
39. $24x^3 - 36x^2 + 6x - 9$ 40. $4x^2 + 25x + 25$
41. $15x^2 + 11x + 2$ 42. $D = \frac{1}{75}x^2 + \frac{1}{2}x + \frac{7}{6}$

Reteaching with Practice

1. $2x^2 + 5x + 3$ **2.** $y^2 - 5y + 6$

3. $6a^2 + a - 2$ **4.** $a^3 + 2a^2 - 5a + 12$

5. $2y^3 + 3y^2 - 9y - 5$

6. $a^3 + 2a^2 - 5a + 12$

7. $2y^3 + 3y^2 - 9y - 5$ **8.** $3x^2 + 11x + 6$

Real-Life Application

1. $60x^2 + 48x + 9$ **2.** $x^2 - 4$

3. $59x^2 + 48x + 13$

4.

x(feet)	6	9	12
Total area of lawn (sq ft)	2457	5301	9225
Area of dog cage (sq ft)	32	77	140
Area of lawn mowed (sq ft)	2425	5224	9085

5. No. For example, the total area of the lawn when $x = 6$ is 2457 square feet. When $x = 12$ the total area of the lawn is 9225 square feet, which is not double 2457 square feet.

Challenge: Skills and Applications

1. $8y^5 - 14y^3 - 15y$

2. $-3a^7 + 21a^4 + 2a^3 - 14$

3. $12c^8 + 4c^4d^2 - d^4$ **4.** $6x^5 + 58x^3 + 80x$

5. $5a^4 - 48a^2b^2 + 27b^4$

6. $x^3 - 5x^2 - 2x + 24$ **7.** $x^3 - x^2 + 16x + 5$

8. $-2x^3 + 9x^2 + 14x - 36$

9. $2x^4 - 9x^3 - 2x^2 - 15$

10. $a = -6, b = -30$ **11.** 5 **12.** -3

13. $(3x - 1), (2x - 1), (x - 1)$

14. $6x^3 - 11x^2 + 6x - 1$ **15.** 231 in.³

Lesson 10.3

Warm-up Exercises

1. $4x^2$ **2.** $64m^2$ **3.** $\frac{1}{16}y^2$ **4.** b^6 **5.** $81n^8$

Daily Homework Quiz

1. $16x - 2x^2 - 6x^3$ **2.** $3t^2 - 4t - 15$

3. $y^2 - 16$ **4.** $x^2 + 9x + 20$ **5.** $6x^2 + x - 2$

Lesson Opener

Allow 15 minutes.

1.

Incr.	Side	New area two factors	New area polynomial
1 in.	$x + 1$	$(x + 1)(x + 1)$	$x^2 + 2x + 1$
2 in.	$x + 2$	$(x + 2)(x + 2)$	$x^2 + 4x + 4$
3 in.	$x + 3$	$(x + 3)(x + 3)$	$x^2 + 6x + 9$
4 in.	$x + 4$	$(x + 4)(x + 4)$	$x^2 + 8x + 16$

2. *Sample answer:* Square first term of binomial, double first term of binomial times last term of binomial, square last term of binomial, add the results.

3.

Incr.	Side	New area two factors	New area polynomial
1 in.	$x - 1$	$(x - 1)(x - 1)$	$x^2 - 2x + 1$
2 in.	$x - 2$	$(x - 2)(x - 2)$	$x^2 - 4x + 4$
3 in.	$x - 3$	$(x - 3)(x - 3)$	$x^2 - 6x + 9$
4 in.	$x - 4$	$(x - 4)(x - 4)$	$x^2 - 8x + 16$

4. *Sample answer:* Square first term of binomial, double first term of binomial times last term of binomial, square last term of binomial, add the results.

Practice A

1. $2xy$ **2.** $2ab$ **3.** $2mn$ **4.** $20x$ **5.** y^2

6. $4b^2$ **7.** $x^2 - 4$ **8.** $t^2 - 9$ **9.** $x^2 - 81$

10. $25 - c^2$ **11.** $n^2 - 25$ **12.** $4x^2 - 49$

13. $49 - d^2$ **14.** $9x^2 - 1$ **15.** $25x^2 - 9$

16. $x^2 + 8x + 16$ **17.** $x^2 - 10x + 25$

18. $x^2 + 16x + 64$ **19.** $4t^2 + 12t + 9$

20. $9y^2 - 30y + 25$ **21.** $16m^2 - 24m + 9$

22. $4m^2 + 16m + 16$ **23.** $4y^2 + 36y + 81$

24. $4k^2 - 12k + 9$ **25.** $w^2 - 25$ **26.** $9z^2 - 1$

27. $x^2 - 4$ **28.** $16x^2 - 9$ **29.** $4x^2 - 81$

30. $25n^2 + 10n + 1$ **31.** $x^2 - 10x + 25$

32. $9x^2 - 12x + 4$ **33.** $49b^2 + 42b + 9$

34. 396 **35.** 891 **36.** 2484 **37.** 75%; 25%

Lesson 10.3 *continued*

Practice B

1. $2mn$ **2.** $2xy$ **3.** $12ab$ **4.** $20x$ **5.** 49

6. $4c^2$ **7.** $x^2 - 9$ **8.** $t^2 - 49$ **9.** $4x^2 - 1$

10. $16x^2 - 9$ **11.** $9n^2 - 9$ **12.** $25x^2 - 4$

13. $9 - 4d^2$ **14.** $49x^2 - 25$ **15.** $x^2 - y^2$

16. $25x^2 - y^2$ **17.** $x^2 - 16y^2$ **18.** $4x^2 - 9y^2$

19. $x^2 + 10x + 25$ **20.** $x^2 - 12x + 36$

21. $x^2 + 18x + 81$ **22.** $4t^2 + 4t + 1$

23. $16y^2 - 8y + 1$ **24.** $m^2 + 14m + 49$

25. $m^2 - 4m + 4$ **26.** $9y^2 - 24y + 16$

27. $9k^2 + 48k + 64$ **28.** $x^2 - 6x + 9$

29. $25t^2 - 20t + 4$ **30.** $16n^2 + 40n + 25$

31. $k^2 - 49$ **32.** $9m^2 - 25$ **33.** $p^2 - q^2$

34. $\frac{1}{4}x^2 - 16$ **35.** $81x^2 - 49$

36. $36n^2 + 36n + 9$ **37.** $y^2 - 2xy + x^2$

38. $25b^2 - 70b + 49$ **39.** $49x^2 + 14x + 1$

40. 1591 **41.** 1575 **42.** 6396

43. $T = 4t^2 - 9$; $91,000$

Practice C

1. $x^2 - 25$ **2.** $t^2 - 16$ **3.** $9x^2 - 25$

4. $49x^2 - 36$ **5.** $25n^2 - 25$ **6.** $81x^2 - 9$

7. $36 - 16d^2$ **8.** $64x^2 - \frac{25}{4}$ **9.** $m^2 - n^2$

10. $\frac{1}{9}x^2 - y^2$ **11.** $x^2 - \frac{16}{25}y^2$ **12.** $36x^2 - 25y^2$

13. $x^2 + 6x + 9$ **14.** $x^2 - 18x + 81$

15. $x^2 + 4xy + 4y^2$ **16.** $4m^2 + 12m + 9$

17. $49y^2 - 42y + 9$ **18.** $25y^2 - 30y + 9$

19. $b^2 - \frac{4}{3}b + \frac{4}{9}$ **20.** $m^2 + m + \frac{1}{4}$

21. $64k^2 + 48k + 9$ **22.** $x^2 - 0.6x + 0.09$

23. $49c^2 - 28cd + 4d^2$

24. $25n^2 + 40mn + 16m^2$ **25.** $c^2 - 25$

26. $25m^2 - 4$ **27.** $v^2 - w^2$ **28.** $\frac{1}{9}x^2 - 36$

29. $81x^2 - 49$ **30.** $64n^2 + 80n + 25$

31. $9y^2 - 12xy + 4x^2$ **32.** $25b^2 - 70bc + 49c^2$

33. $49x^2 + 14xy + y^2$ **34.** 1584 **35.** 324

36. 2704 **37.** $8x$; 40 in.^{2}; 48 in.^{2}; 56 in.^{2}

38. $4x + 16$; 36 in.^{2}; 44 in.^{2}; 52 in.^{2}

Reteaching with Practice

1. $x^2 - 25$ **2.** $9x^2 - 4$ **3.** $x^2 - 4y^2$

4. $m^2 + 2mn + n^2$ **5.** $9x^2 - 12x + 4$

6. $49y^2 + 28y + 4$ **7.** $x^2 + 10x + 25$

8. $x^2 - 9$

Cooperative Learning Activity

1. The area of the large square is $a \cdot a$, which equals a^2; similarly, the area of the smaller square equals b^2. **2.** The area of the square with sides of length $(a + b)$ equals $(a + b)^2$. The area of the square with sides of length $(a - b)$ equals $(a - b)^2$. **3.** $2(a^2 + b^2) = (a + b)^2 + (a - b)^2$

Interdisciplinary Application

1.

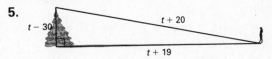

2. $t^2 - 16t + 28 = 0$

3. $14^2 - 16(14) + 28 \overset{?}{=} 0$

$196 - 224 + 28 \overset{?}{=} 0$

$0 \overset{\checkmark}{=} 0$

and

$2^2 - 16(2) + 28 \overset{?}{=} 0$

$4 - 32 + 28 \overset{?}{=} 0$

$0 \overset{\checkmark}{=} 0$

4. When $t = 14$, the lengths are 7, 24, and 25. When $t = 2$, the lengths are -5, 12, and 13. Because the length cannot be negative, $t = 2$ is not a correct answer.

5.

6. $t^2 - 62t + 861 = 0$

7. $21^2 - 62(21) + 861 \overset{?}{=} 0$

$441 - 1302 + 861 \overset{?}{=} 0$

$0 \overset{\checkmark}{=} 0$

and

$41^2 - 62(41) + 861 \overset{?}{=} 0$

$1681 - 2542 + 861 \overset{?}{=} 0$

$0 \overset{\checkmark}{=} 0$

Lesson 10.3 *continued*

8. When $t = 21$, the height of the tree is -9 feet. Because this is not possible, $t = 21$ is not a correct answer. When $t = 41$, height $= 11$ feet, length $= 60$ feet and distance $= 61$ feet.

Math and History

1. 25%; 75%, 3 to 1   **2.** 3 to 1

3. accept reasonable responses

Challenge: Skills and Applications

1. $2x^2 + 4x + 34$ **2.** $-4x - 20$

3. $3x^2 - 10x - 8$ **4.** $-19x^2 - 7x - 67$

5. $(a + b + c)^2 = (a + b + c)(a + b + c) =$
$a(a + b + c) + b(a + b + c) +$
$c(a + b + c) = a^2 + ab + ac + ab + b^2 +$
$bc + ac + bc + c^2 = a^2 + b^2 + c^2 +$
$2ab + 2ac + 2bc$

6. $x^2 + 9y^2 + 4z^2 - 6xy + 4xz - 12yz$

7. $x^4 + 8x^3 + 6x^2 - 40x + 25$

8. a. $x^3 - 125$ **b.** $x^3 - 8$
c. $(x - a)(x^2 + ax + a^2) = x^3 - a^3$

9. a. $x^4 - 625$
b. $(x - a)(x^3 + ax^2 + a^2x + a^3) = x^4 - a^4$

10. $(2)(3)(4)(5) + 1 = 121 = 11^2$

11. $(3)(4)(5)(6) + 1 = 361 = 19^2$

12. $n^4 + 2n^3 - n^2 - 2n + 1$

13. Using the results from Exercise 5 you get the following:
$(n^2 + n - 1)^2 = (n^2)^2 + n^2 + (-1)^2 +$
$2(n^2)(n) + 2(n^2)(-1) + 2(n)(-1) =$
$n^4 + n^2 + 1 + 2n^3 - 2n^2 - 2n =$
$n^4 + 2n^3 - n^2 - 2n + 1$

Quiz 1

1. $3n^2 + 8n + 1$ **2.** $-1y^3 + 8y^2 - 3y + 1$

3. $-9t^4 + 12t^3 - 15t^2$ **4.** $8y^2 - 2y - 6$

5. $m^2 - 8m + 16$ **6.** $9a^2 - 25$

7. $(y^2 + 22y + 120)$ ft^2

Lesson 10.4

Warm-up Exercises

1. ± 3 **2.** ± 5 **3.** ± 8 **4.** $2x^2 - 7x + 3$

5. $x^2 + 16x + 64$

Daily Homework Quiz

1. $25x^2 - 25$ **2.** $x^2 - 22x + 121$

3. $9n^2 + 24mn + 16m^2$

4. $\left(\frac{1}{2}H + \frac{1}{2}T\right)^2 = \frac{1}{4}H^2 + \frac{1}{2}HT + \frac{1}{4}T^2$

Lesson Opener

Allow 10 minutes.

1. *Sample answer:*
When $(2 + 0.1x)(100 - x) = 0$, you are not collecting any money.   **2.** Yes; $0(100 - x) = 0$

3. Yes; $(2 + 0.1x)0 = 0$ **4.** $-20, 100$

5. *Sample answer:* When
$(4 + 0.5y)(50 - 2y) = 0$, you are not collecting any money.   **6.** Yes; $0(50 - 0.5y) = 0$

7. Yes; $(4 + 2y)0 = 0$ **8.** $-8, 25$

Practice A

1. no   **2.** no   **3.** yes   **4.** yes   **5.** no   **6.** yes

7. $-2, -5$ **8.** $-3, 3$ **9.** $-1, 5$ **10.** $-5, -3$

11. $3, 5$ **12.** $-7, 7$ **13.** $-4, -8$ **14.** $-1, 1$

15. $-5, 3$ **16.** $-\frac{1}{4}, -\frac{3}{8}$ **17.** $-1.2, 7.1$

18. -5 **19.** $-3, -5$ **20.** $\frac{1}{3}, -4$ **21.** $-3, 5$

22. $4, -5$ **23.** $\frac{7}{2}, 2$ **24.** $-2, -5$ **25.** $-\frac{4}{3}, 3$

26. $-3, -2$ **27.** $\frac{1}{2}, \frac{3}{4}$ **28.** $-\frac{3}{4}, \frac{3}{4}$ **29.** $\frac{9}{2}, -4$

30. $-\frac{1}{5}, 2$ **31.** B   **32.** A   **33.** C

34. at $t = 10$ sec; use zero-product property

Practice B

1. $-1, -6$ **2.** $-4, 4$ **3.** $-9, 8$ **4.** $-7, -2$

5. $8, 9$ **6.** $-4.2, 4.2$ **7.** $-\frac{3}{4}, -\frac{5}{8}$ **8.** $-\frac{1}{2}, \frac{1}{2}$

9. $-5.4, 3$ **10.** -4 **11.** $3.2, \frac{3}{2}$ **12.** $6, -6$

13. $-4, -7$ **14.** $\frac{1}{5}, -2$ **15.** $-4, 6$ **16.** $\frac{7}{2}, -4$

17. $\frac{5}{2}, \frac{5}{3}$ **18.** $-3, -9$ **19.** $-\frac{7}{5}, \frac{15}{7}$ **20.** $-4, -2$

21. $3.2, 2.1$ **22.** $-3.1, 3.2$ **23.** $\frac{1}{4}, -\frac{1}{4}$

24. $-\frac{1}{15}, \frac{1}{6}$ **25.** B   **26.** A   **27.** C

28.

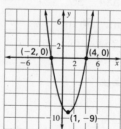

29.

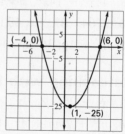

Lesson 10.4 *continued*

30.

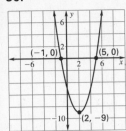

31.

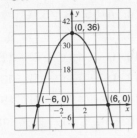

32.

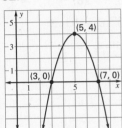

33.

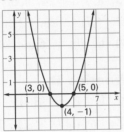

34. at $t = 2$ sec; use zero-product property

Practice C

1. $-2, -7$ **2.** $-9, 9$ **3.** $-4, 8$ **4.** $-5, -1$

5. $9, 3$ **6.** $-7.2, 7.2$ **7.** $-\frac{3}{5}, \frac{3}{5}$ **8.** $-\frac{2}{3}, \frac{2}{3}$

9. $-6.1, 5.3$ **10.** -6 **11.** $5.2, \frac{5}{2}$ **12.** $4, -4$

13. $-3, -6$ **14.** $\frac{3}{5}, -\frac{5}{2}$ **15.** $-3, 4$ **16.** $\frac{2}{7}, -3$

17. $3.9, 1.4$ **18.** $-\frac{1}{3}, -\frac{1}{9}$ **19.** $-0.8, 2.7$

20. $-2, -1, 4$ **21.** $3, 2, -1$ **22.** $-2.3, 4.1$

23. $\frac{1}{6}, -\frac{1}{6}$ **24.** $-\frac{1}{12}, \frac{1}{18}$ **25.** B **26.** A **27.** C

28.

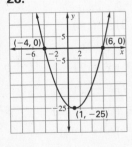

29.

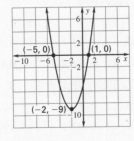

30.

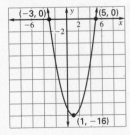

31.

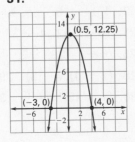

32.

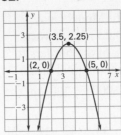

33.

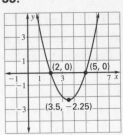

34. Use the zero-product property to find
x-intercepts and find the distance between them,
35 ft; $(0, 45.9375)$

Reteaching with Practice

1. $6, -6$ **2.** $5, 1$ **3.** $-4, -3$ **4.** 5 **5.** -3

6. -2 **7.** $-3; -1; (-2, -1)$

8. $2; 4; (3, -1)$ **9.** $1; -5; (-2, -9)$

Real-Life Application

1.

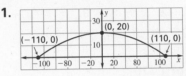

2. 220 feet

3. 20 feet **4.**

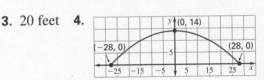

5. horizontal distance $= 56$ feet;
height $= 14$ feet **6.** Answers will vary.

Challenge: Skills and Applications

1. $x(x - 3)(x + 2) = 0$ **2.** $x^2(x + 4) = 0$

3–6. Accept equivalent equations.

3. $\left(x + \frac{1}{2}\right)(x - 5)(x - 7) = 0$, or
$(2x + 1)(x - 5)(x - 7) = 0$

4. $\left(x - \frac{2}{3}\right)(x - 4)\left(x + \frac{5}{6}\right) = 0$, or
$(3x - 2)(x - 4)(6x + 5) = 0$

5. $(x + k)\left(x - \frac{k}{5}\right)(x - 4k) = 0$, or
$(x + k)(5x + k)(x - 4k) = 0$

6. $(x - 2k)\left(x - \frac{4k}{7}\right)\left(x + \frac{5}{9k}\right) = 0$, or
$(x - 2k)(7x - 4k)(9kx + 5) = 0$

Lesson 10.4 *continued*

7. $(12 - 2x)(20 - 4x) = 126$ (Models may vary. For example, a student could instead write a model based on adding the area of the borders to the area of the printed part to get the total area.)

8. 1.5 in., 3 in.   **9.** $12 + n$, $6000 - 200n$, $(12 + n)(6000 - 200n)$

10.

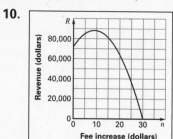

11. $42   **12.** $21

Lesson 10.5

Warm-up Exercises

1. $x^2 + 8x + 12$ **2.** $x^2 - 81$ **3.** $x^2 - 6x + 9$
4. $2x^2 + 5x - 25$ **5.** $x^2 - \frac{1}{16}$

Daily Homework Quiz

1. 7   **2.** $-3, 9$ **3.** $1, -\frac{5}{3}$
4. $-1, 1, (0, -1)$

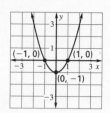

Lesson Opener

Allow 10 minutes.

1. $x^2 + 5x + 6$ **2.** Add 3 and 2.
3. Multiply 3 and 2.   **4.** $x^2 - 3x - 4$
5. Add -4 and 1.   **6.** Multiply -4 and 1.
7. $x^2 + 3x - 18$ **8.** Add 6 and -3.
9. Multiply 6 and -3.   **10.** $x^2 - 8x + 15$
11. Add -3 and -5.   **12.** Multiply -3 and -5.

13. *Sample answer:* The last term of the product of two binomials is the product of the last two terms of the binomials. The coefficient of the middle term of the product of two binomials is the sum of the last two terms of the binomials.

Practice A

1. $(x + 3)(x + 2)$ **2.** $(x + 4)(x + 3)$ **3.** D
4. A   **5.** B   **6.** C   **7.** $(x + 2)(x + 4)$
8. $(x + 4)(x - 1)$ **9.** $(x + 1)(x + 2)$
10. $(x - 4)(x + 2)$ **11.** $(x + 4)(x + 3)$
12. $(x - 1)(x - 5)$ **13.** $(x + 5)(x - 4)$
14. $(x + 4)(x + 4)$ **15.** $(x - 6)(x - 4)$
16. $-4, 1$ **17.** $2, 3$ **18.** $-6, 3$ **19.** $18, -2$
20. $-1, -7$ **21.** $-5, 2$ **22.** $-7, 2$
23. $8, -1$ **24.** $4, 5$ **25.** $-6, 8$ **26.** $-3, -9$
27. $-7, 4$ **28.** 12 cm   **29.** 6 in. × 10 in.
30. height $= 6$ in.; base $= 9$ in.

Practice B

1. D   **2.** E   **3.** B   **4.** F   **5.** A   **6.** C
7. $(x + 2)(x - 7)$ **8.** $(x - 3)(x - 5)$
9. $(x + 3)(x + 5)$ **10.** $(x - 1)(x - 4)$
11. $(x - 7)(x + 6)$ **12.** $(x + 8)(x - 2)$
13. $(x - 8)(x - 8)$ **14.** $(x + 4)(x + 9)$
15. $(x - 12)(x - 3)$ **16.** $-8, 5$ **17.** $9, 7$
18. $4, 7$ **19.** $7, -1$ **20.** 3   **21.** $-3, -5$
22. $-3, 2$ **23.** $-12, 1$ **24.** $7, -4$ **25.** $-7, 1$
26. $-8, 1$ **27.** $6, -2$ **28.** yes; $(x + 5)(x + 12)$
29. no   **30.** yes; $(x - 9)(x + 4)$
31. yes; $(x + 3)(x + 10)$ **32.** yes; $(x + 5)(x + 6)$
33. no   **34.** $x - 10$ **35.** 14 ft or 6 ft

Practice C

1. $(x + 2)(x + 3)$ **2.** $(x + 2)(x + 4)$
3. $(x - 1)(x - 3)$ **4.** $(x - 5)(x - 6)$
5. $(x - 4)(x + 2)$ **6.** $(x - 4)(x + 3)$
7. $(x + 7)(x - 4)$ **8.** $(x + 7)(x - 2)$
9. $(x + 3)(x + 5)$ **10.** $(x - 10)(x - 10)$
11. $(x + 9)(x + 8)$ **12.** $(x - 16)(x + 4)$
13. $4, 9$ **14.** $10, -7$ **15.** $-9, 5$ **16.** $-4, -7$
17. $4, 11$ **18.** $-6, 3$ **19.** $9, -7$ **20.** $7, -2$
21. $10, 1$ **22.** $4, -3$ **23.** $1, 3$ **24.** $-7, 2$
25. $6, -2$ **26.** $-2, -4$ **27.** $-10, 8$
28. $(x - 4)(x - 8)$ **29.** $(x - 16)(x + 3)$
30. $(x - 10)(x + 9)$ **31.** $(x - 12)(x + 7)$

Lesson 10.5 *continued*

32. $(x - 6)(x - 11)$ **33.** no

34. $x^2 - 17x + 60$

35. $x^2 - 2x - 360$ **36.** $x^2 - 25x = 0$

37. $(x + 12)(x + 6)$ **38.** $-5, -13$

39. 1 ft by 7 ft

Reteaching with Practice

1. $(x + 3)(x + 2)$ **2.** $(x + 1)(x + 5)$

3. $(x + 2)(x + 1)$ **4.** $(x - 2)(x - 1)$

5. $(x - 3)(x - 4)$ **6.** $(x - 2)(x - 3)$

7. $(x - 2)(x + 1)$ **8.** $(x - 6)(x + 2)$

9. $(x - 4)(x + 2)$ **10.** $-3, -5$

11. 2, 6   **12.** 1, -4

Real-Life Application

1. $(x + 30)(x - 10)$ **2.** $x^2 + 20x - 300$

3. $x = 50$ **4.** 21 rows   **5.** No; 640 square feet

Challenge: Skills and Applications

1. $(x^{1/2} - 5)(x^{1/2} - 2)$ **2.** $(x^3 - 6)(x^3 + 2)$

3. $(\sqrt{x} - 9)(\sqrt{x} + 5)$

4. $(\sqrt{x} + 1)(\sqrt{x} + 10)$ **5.** $\left(\dfrac{1}{x} - 7\right)\left(\dfrac{1}{x} - 8\right)$

6. $\left(\dfrac{1}{x} - 9\right)\left(\dfrac{1}{x} + 12\right)$ **7.** $2, -2, 3, -3$

8. $2\sqrt{2}, -2\sqrt{2}, \sqrt{3}, \sqrt{3}$ **9.** $-2, 3$

10. $-1, -\frac{3}{10}$ **11.** 2; 4, -4

12. $(x - 6.6)(x - 7)$ **13.** 3.3 ft, 3.5 ft

14. 60 ft by 60 ft

Lesson 10.6

Warm-up Exercises

1. $(x - 6)(x - 4)$ **2.** $(x + 9)(x + 1)$

3. $(x + 10)(x - 4)$ **4.** $(x - 10)(x + 4)$

5. $(x - 8)(x - 3)$

Daily Homework Quiz

1. $(y + 7)(y - 2)$ **2.** $(x + 8)(x - 11)$

3. $(t + 18)(t + 5)$ **4.** $-4, 8$

5. yes; $(y + 12)(y - 10)$

Lesson Opener

Allow 10 minutes.

1. Check work.   **2.** *Sample answer:* The graphs are different.   **3.** $2x^2$; they are the same.

4. 3; they are the same.

5. $5x$; they are different.   **6.** *Sample answer:* The graphs are the same.   **7.** $2x^2$; they are the same.   **8.** 3; they are the same.   **9.** $7x$; they are the same.   **10.** The product in Question 6 is the same as the trinomial in Question 1. The product in Question 2 is different than the trinomial in Question 1.

Practice A

1. $(2x + 2)(x + 1)$ **2.** $(2x + 3)(x + 2)$ **3.** C

4. B   **5.** D   **6.** A   **7.** A   **8.** $(3x - 1)(x + 4)$

9. $3(x - 2)(x - 2)$ **10.** $(2x + 1)(x - 3)$

11. $(3x - 2)(x + 4)$ **12.** $(7x - 3)(x - 4)$

13. not factorable   **14.** $(5x + 2)(x + 1)$

15. $(2x - 3)(3x - 1)$ **16.** $(5x + 1)(6x - 1)$

17. not factorable   **18.** $(2x + 1)(x - 5)$

19. $1, -4$ **20.** $\frac{2}{3}, -5$ **21.** $-\frac{5}{2}, 1$ **22.** $\frac{3}{5}, 1$

23. $-\frac{5}{3}, -3$ **24.** $\frac{1}{2}, \frac{3}{2}$ **25.** $\frac{10}{7}, -3$ **26.** $-\frac{3}{5}, 5$

27. $\frac{7}{2}, 4$ **28.** 1.25 sec

Practice B

1. B   **2.** C   **3.** F   **4.** E   **5.** A   **6.** D

7. $2(x + 4)(x - 2)$ **8.** $(5x - 2)(x - 3)$ **9.** A

10. $(2x + 5)(x + 3)$ **11.** not factorable

12. $(5x - 1)(2x + 3)$ **13.** $(5x + 1)(2x + 3)$

14. $(4x + 3)(2x - 1)$ **15.** not factorable

16. $(6x - 1)(2x + 3)$ **17.** not factorable

18. $(5x + 3)(2x - 3)$ **19.** $2, -\frac{1}{3}$ **20.** $3, \frac{3}{2}$

21. $-\frac{2}{3}, -1$ **22.** $-\frac{1}{2}, -\frac{3}{4}$ **23.** $-\frac{1}{2}, \frac{5}{2}$ **24.** $\frac{3}{4}, -\frac{1}{3}$

25. $-\frac{5}{3}, \frac{3}{5}$ **26.** $\frac{1}{4}, \frac{5}{2}$ **27.** $-\frac{1}{3}, -\frac{1}{2}$

28. No; the cost of the T-shirts decreased by $1 each week.   **29.** 3 sec

Practice C

1. A   **2.** $(2x - 5)(3x + 1)$ **3.** B

4. $(2x - 7)(x + 3)$ **5.** not factorable

6. $(3x + 1)^2$ **7.** $(3x + 5)(x + 2)$

Answers

Lesson 10.6 *continued*

8. $(2x + 3)(x - 2)$ **9.** not factorable
10. $(7x + 8)(2x - 5)$ **11.** not factorable
12. $6(x - 3)^2$ **13.** $-\frac{1}{2}, -3$ **14.** $-5, \frac{1}{3}$
15. $\frac{1}{3}, -4$ **16.** $-\frac{1}{2}, -\frac{5}{3}$ **17.** $-\frac{1}{3}, -2$
18. $-\frac{1}{3}, \frac{3}{4}$ **19.** $-\frac{1}{2}, -1$ **20.** $-\frac{1}{6}, -\frac{5}{2}$
21. $-\frac{3}{7}, -2$ **22.** $\frac{3}{2}, -\frac{3}{2}$ **23.** $0, -6$
24. $2 + \sqrt{3}, 2 - \sqrt{3}$ **25.** $3, 7$ **26.** -3
27. $-\frac{1}{3}, \frac{1}{4}$ **28.** $\dfrac{-3 + \sqrt{33}}{4}, \dfrac{-3 - \sqrt{33}}{4}$
29. $4, -4$ **30.** $-\frac{5}{6}, \frac{7}{3}$ **31.** 3 sec   **32.** 2.5 sec

Reteaching with Practice

1. $(5x + 1)(x + 2)$ **2.** $(2x + 3)(x + 1)$
3. $(3x + 7)(x + 1)$ **4.** $(9x + 2)(x + 7)$
5. $(6x - 5)(x - 3)$ **6.** $(4x + 1)(2x + 9)$
7. $-3, -\frac{1}{2}$ **8.** $3, \frac{2}{5}$ **9.** $\frac{2}{3}, -\frac{1}{2}$

Interdisciplinary Application

1. $(x - 4)(8x - 3)$ **2.** $(9x - 8)(10x - 7)$
3. $(5x - 6)(2x - 9)$ **4.** $(3x - 10)(4x - 1)$
5. $(6x - 11)(7x - 2)$
6. $(11x - 12)(12x - 5)$

Letter	A	B	C	D	E	F
Code Number	1	−4	8	−3	9	−8
Letter	M	N	O	P	Q	R
Code Number	3	−10	4	−1	6	−11

Letter	G	H	I	J	K	L
Code Number	10	−7	2	−9	5	−6
Letter	S	T	U	V	Y	Z
Code Number	7	−2	11	−12	12	−5

7. PATTERNED VESSEL
8. CARVED STOOL
9. MASAI NECKLACE
10. IVORY MASK

Challenge: Skills and Applications

1. $(x - 2y)(x - 7y)$ **2.** $(3a - 4b)(2a + b)$
3. $(4p - q)(3p + 5q)$ **4.** $(2ab + 5)(ab - 4)$
5. $(x - 4)(4x + 23)$ **6.** $(3\sqrt{x} - 1)(5\sqrt{x} + 2)$
7. $2, -1$ **8.** $\frac{3}{2}, -\frac{3}{2}, 1, -1$ **9.** $\frac{1}{2}, 5$ **10.** $3, \frac{2}{5}$
11. Accept equivalent forms.
$(x + 2)^3 - x^3 = 296$
12. $6x^2 + 12x - 288 = 0$ **13.** 6 in.

Quiz 2

1. $x = -2, x = 6$
2. $x = -3, x = 1$;
vertex $(-1, -4)$

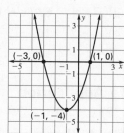

3. $(x - 5)(x - 2)$ **4.** $-4, 5$
5. $(4x + 9)(x - 1)$ **6.** $\frac{7}{5}, -3$

Lesson 10.7

Warm-up Exercises

1. $4n^2 + 12n + 9$ **2.** $p^2 - 12p + 36$
3. $4y^2 - 1$ **4.** $(x - 5)(x + 1)$
5. $(x + 1)(x + 1)$

Daily Homework Quiz

1. $2(7x - 3)(x + 3)$ **2.** not factorable
3. $4(n + 9)(n - 8)$ **4.** $\frac{4}{3}, -\frac{9}{2}$
5. $-5, 0.6$ **6.** $-\frac{1}{2}, 3$

Lesson Opener

Allow 10 minutes.
1. $x^2 - 1$ **2.** $m^2 - 4$ **3.** $y^2 - 16$
4. $n^2 - 64$ **5.** They are all binomials and both terms are perfect squares.   **6.** The first term of the answer is the square of the first term of the binomial.   **7.** The second term of the answer is the square of the second term of the binomial.
8. $x^2 + 2x + 1$ **9.** $m^2 + 4m + 4$
10. $y^2 - 8y + 16$ **11.** $n^2 - 16n + 64$

Lesson 10.7 *continued*

12. They are all trinomials and the first and last terms are perfect squares.   **13.** The first term of the answer is the square of the first term of the binomial.   **14.** The last term of the answer is the square of the second term of the binomial.

15. They are the same.

Graphing Calculator Activity

1. -1 **2.** $2, 5$ **3.** $2, 3$ **4.** $-3, 2$ **5.** -1

6. 3 **7.** Answers will vary. *Sample answer:* A perfect square trinomial (a trinomial that factors into two identical binomials) will only have one solution.

Practice A

1. $3(x^2 + 2x + 1)$ **2.** $2(x^2 - 4x + 4)$

3. $4(x^2 + 2x + 1)$ **4.** $3(x^2 - 4x + 4)$

5. $5(x^2 + 6x + 9)$ **6.** $-3(x^2 + 6x + 9)$

7. C   **8.** B   **9.** F   **10.** A   **11.** D   **12.** E

13. $(x + 5)(x - 5)$ **14.** $(x + 7)(x - 7)$

15. $(2x + 3)(2x - 3)$ **16.** $\left(x + \frac{1}{2}\right)\left(x - \frac{1}{2}\right)$

17. $(x + y)(x - y)$ **18.** $(5x + 4)(5x - 4)$

19. $(3x + 4)(3x - 4)$ **20.** $(4 + x)(4 - x)$

21. $2(10x + 3)(10x - 3)$ **22.** $(x + 5)^2$

23. $(x - 9)^2$ **24.** $(2x + 1)^2$ **25.** $\left(x + \frac{1}{2}\right)^2$

26. $(x - y)^2$ **27.** $(2x + 7)^2$ **28.** $(5x - 2)^2$

29. $(3x - 4y)^2$ **30.** $5(2x + 3)^2$ **31.** $3, -3$

32. -2 **33.** $\frac{2}{3}, -\frac{2}{3}$ **34.** 4 **35.** $2, -2$ **36.** 5

37. $\pi x^2 - \pi y^2$; $\pi(x - y)(x + y)$; 21π cm^2

Practice B

1. E   **2.** B   **3.** A   **4.** F   **5.** C   **6.** D

7. $(x + 3)(x - 3)$ **8.** $(x + 12)(x - 12)$

9. $\left(x + \frac{1}{3}\right)\left(x - \frac{1}{3}\right)$ **10.** $(x + 0.4)(x - 0.4)$

11. $(2x + 7)(2x - 7)$ **12.** $(3x + 5)(3x - 5)$

13. $(p + q)(p - q)$ **14.** $(8x + 3y)(8x - 3y)$

15. $(10 - x)(10 + x)$ **16.** $(7 + x)(7 - x)$

17. $(2a + b)(2a - b)$ **18.** $12(x + 2)(x - 2)$

19. $(x + 9)^2$ **20.** $(x - 4)^2$ **21.** $(2x + 1)^2$

22. $(5x - 3)^2$ **23.** $(a - b)^2$ **24.** $(x + 2y)^2$

25. $\left(2x + \frac{1}{2}\right)^2$ **26.** $(3x - 0.1)^2$ **27.** $2(2x + y)^2$

28. $3(x - 5y)^2$ **29.** $(5 - 2x)^2$ **30.** $4(2x + 3)^2$

31. $4, -4$ **32.** $\frac{5}{2}$ **33.** $\frac{1}{2}, -\frac{1}{2}$ **34.** $\frac{4}{3}$ **35.** $2, -2$

36. $-\frac{5}{2}$

37. $81x^2 + 36x + 4$, $(9x + 2)^2$, $6\frac{4}{9}$ in.

Practice C

1. $(x + 3)(x - 3)$ **2.** $(11 + x)(11 - x)$

3. $(3 + 2x)(3 - 2x)$ **4.** $(3x + 7)(3x - 7)$

5. $(5x + 2y)(5x - 2y)$ **6.** $\left(7x + \frac{1}{2}\right)\left(7x - \frac{1}{2}\right)$

7. $\left(\frac{1}{2}x + \frac{1}{3}y\right)\left(\frac{1}{2}x - \frac{1}{3}y\right)$ **8.** $(5x + 0.2)(5x - 0.2)$

9. $(3x + 7y)(3x - 7y)$ **10.** $2(3x + 4)(3x - 4)$

11. $5\left(2x + \frac{1}{2}\right)\left(2x - \frac{1}{2}\right)$ **12.** $-7(x + 2)(x - 2)$

13. $-3(3x + 2y)(3x - 2y)$

14. $(x^2 + 4)(x + 2)(x - 2)$

15. $(2x + 0.3)(2x - 0.3)$ **16.** $(x + 7)^2$

17. $(2x + y)^2$ **18.** $(3x - 2)^2$ **19.** $(5 - x)^2$

20. $(3 + 2x)^2$ **21.** $(3x - 2y)^2$ **22.** $\left(x + \frac{3}{4}\right)^2$

23. $(2a - 5b)^2$ **24.** $3(2x + 7)^2$ **25.** $-4(x - 3)^2$

26. $(5x - 0.2)^2$ **27.** $8(x + 3)^2$ **28.** $-5(2x + 3)^2$

29. $(7 - 3x)^2$ **30.** $(x^2 + y^2)^2$ **31.** $\frac{5}{3}, -\frac{5}{3}$

32. 4 **33.** $\frac{1}{3}, -\frac{1}{3}$ **34.** $\frac{1}{2}$ **35.** $6, -6$ **36.** $\frac{3}{2}$

37. $A = b_1 h + \frac{1}{2}(b_2 - b_1)h = h\left[b_1 + \frac{1}{2}b_2 - \frac{1}{2}b_1\right]$
$= h\left[\frac{1}{2}b_1 + \frac{1}{2}b_2\right] = \frac{1}{2}h\left[b_1 + b_2\right]$

38. $4x^2 - 9$, $(2x + 3)(2x - 3)$, 391

Reteaching with Practice

1. $(4 + 3y)(4 - 3y)$ **2.** $(2q + 7)(2q - 7)$

3. $(6 + 5x)(6 - 5x)$ **4.** $(x - 9)^2$

5. $(2n + 5)^2$ **6.** $(4y + 1)^2$ **7.** 10

8. $\frac{1}{2}$ **9.** 4 **10.** ± 7 **11.** $\pm\frac{8}{3}$ **12.** $\pm\frac{9}{2}$

Interdisciplinary Application

1. $A = \pi R^2$ **2.** $A = \pi r^2$ **3.** $A = \pi(R^2 - r^2)$

4. $A = \pi(R + r)(R - r)$ **5.** $A = 4R^2$

6. $W = 4R^2 - \pi R^2 + \pi r^2$ **7.** $A = .98175$

8. $W = 1.26825$ **9.** $\dfrac{1.26825}{2.25} = 56\%$ waste

Challenge: Skills and Applications

1. $(7rs - 10t)(7rs + 10t)$; difference of two squares pattern   **2.** $(b + 10)(b - 4)$; difference of two squares pattern   **3.** $(x + 5)^2$; perfect square trinomial pattern   **4.** $(r - 3)^2(r + 3)^2$;

Lesson 10.7 *continued*

perfect square trinomial pattern and the difference of two squares pattern

5. $(k + 2)(k - 2)(k^2 + 4)$; difference of two squares pattern   **6.** $(2 - n)(2 + n)(6 + n^2)$; difference of two squares pattern   **7.** $\frac{3}{2}, -\frac{3}{2}$ **8.** $5, -9$ **9.** $\sqrt{6}, -\sqrt{6}$ **10.** 4   **11. a.** $x^4 - 81$

b. $x^4 - a^4 = (x - a)(x^3 + ax^2 + a^2x + a^3)$

c. yes;
$x^n - a^n = (x - a)(x^{n-1} + ax^{n-2} + \ldots + a^{n-1})$

12. a. $(x^2 + 8)^2 - (4x)^2$

b. $(x^2 - 4x + 8)(x^2 + 4x + 8)$

c. $(x^2 - 8x + 32)(x^2 + 8x + 32)$

Lesson 10.8

Warm-up Exercises

1. 3   **2.** 17   **3.** 15   **4.** $12x^2 + 27x$
5. $-6x^3 - 15$

Daily Homework Quiz

1. $36(x + 2)(x - 2)$ **2.** $(r + 9)^2$
3. $2(2y - 5)^2$ **4.** $-\frac{3}{2}, \frac{3}{2}$ **5.** -6
6. $\dfrac{\sqrt{3}}{2}$ sec, or about 0.87 sec

Lesson Opener

Allow 10 minutes.
ACROSS   1. $2x^2$ **2.** $4x^3$ **3.** $5x$ **4.** x **6.** $6x^2$
7. $12x$ **9.** $80x$
DOWN   1. $2x$ **2.** $4x$ **3.** $5x$ **4.** x **5.** $3x^2$
6. $6x$ **8.** $20x$

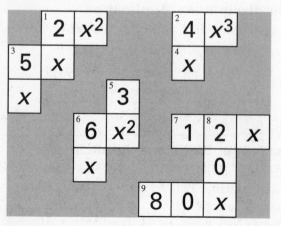

Practice A

1. $3(x + 6)$ **2.** $-2(c - 5)$ **3.** $4(y^2 + y + 2)$
4. $2x(3x^2 - x + 1)$ **5.** $d^2(d^2 + d - 2)$
6. $2a(5a^2 - 6a + 2)$ **7.** yes   **8.** no; $2(n + 2)^2$
9. no; $3(x - 1)(x + 1)$ **10.** yes
11. no; $2(x + 2)(x + 3)$ **12.** yes
13. $13(x + 1)$ **14.** $(c + 2)(c - 2)$
15. $(m - 5)(m + 3)$ **16.** $(2x + 7)(x + 4)$
17. $14x(1 - 2x)$ **18.** $15m(3n - 2m)$
19. $2(x + 7)(x + 1)$ **20.** $5(x - 3)(x + 3)$
21. $7(2t + 1)(t - 3)$ **22.** $x(x + 3)^2$
23. $(x^2 + 2)(x + 1)$ **24.** $(d^2 + 3)(d + 1)$
25. $-1, -3$; *Sample answer:* factoring
26. $4, -4$; *Sample answer:* factoring
27. no solution
28. $2 + \sqrt{6}, 2 - \sqrt{6}$; *Sample answer:* quadratic formula
29. no solution; *Sample answer:* quadratic formula
30. $\dfrac{-1 + \sqrt{5}}{4}, \dfrac{-1 - \sqrt{5}}{4}$; *Sample answer:* quadratic formula
31. $4(x + 2)(x + 1)$ **32.** $(x + 2)$ by $(y - 4)$

Practice B

1. $2x(3x + 5)$ **2.** $5c(c^2 - 5c + 2)$
3. $3y(5y^2 + 2y - 7)$ **4.** $2x^2(5x^2 + 8x + 2)$
5. $d^2(4d^2 + d - 3)$ **6.** $2a^2(4a^3 - 5a + 9)$
7. yes   **8.** no; $5(n + 4)^2$
9. no; $2(x - 2)(x + 2)$ **10.** yes
11. no; $2(x + 1)(x + 6)$ **12.** yes
13. $6x^2(x + 3)$ **14.** $3c(c + 2)(c - 2)$
15. $-2m(5m^2 + 1)$ **16.** $7a^2(5a - 4)$
17. $16x(2 - 3x)$ **18.** $5x(7y - 12x)$
19. $3(m + 6)(m + 2)$ **20.** $4(x - 4)(x + 5)$
21. $2t(t + 3)(t - 2)$ **22.** $6x(x + 2)^2$
23. $(x^2 + 4)(x + 1)$ **24.** $(d^2 + 3)(d + 2)$
25. $-1, -6$; *Sample answer:* factoring
26. no solution   **27.** $\frac{7}{2}$; *Sample answer:* factoring
28. $\dfrac{3 + \sqrt{3}}{3}, \dfrac{3 - \sqrt{3}}{3}$; *Sample answer:* quadratic formula

Lesson 10.8 *continued*

29. no solution

30. $\dfrac{-2 + \sqrt{19}}{5}, \dfrac{-2 - \sqrt{19}}{5}$; *Sample answer:* quadratic formula

31. 2 sec   **32.** 4 sec

Practice C

1. $3x(x - 4)$ **2.** $4c(c^2 - 3c + 2)$

3. $-7y(y^2 - 5y + 1)$ **4.** $5x\left(\frac{2}{3}x^2 + \frac{1}{3}x + 7\right)$

5. $3d^2(5d^2 - 2d + 1)$ **6.** $8ab(a^3 + 6a - 11)$

7. yes   **8.** no; $2n(2n + 1)(n - 5)$

9. no; $7x(3x - 5)(3x + 5)$ **10.** yes

11. no; $8(3x - 7)(2x + 4)$ **12.** yes

13. $3x(7x - 5)$ **14.** $-4c^2(c - 3)$

15. $5m(m + 5)^2$ **16.** $2y(3y - 5)(y + 2)$

17. $3t(2t + 5)(t - 1)$ **18.** $7x(4 - 3x)(2 + x)$

19. $(x^2 + 3)(x - 2)$ **20.** $5x(x + 2)(x - 2)$

21. $(t - 2)(t + 2)(t + 3)$

22. $(x + 1)(x - 1)(2x + 3)$ **23.** $(x^2 + 3)(x - 4)$

24. $(2d^2 + 3)(d - 5)$

25. $3, -\frac{2}{7}$; *Sample answer:* factoring

26. no solution

27. $\dfrac{-3 + \sqrt{15}}{2}, \dfrac{-3 - \sqrt{15}}{2}$; *Sample answer:* quadratic formula

28. no solution

29. $\dfrac{5 + \sqrt{37}}{6}, \dfrac{5 - \sqrt{37}}{6}$; *Sample answer:* quadratic formula

30. $\frac{9}{5}, 2$; *Sample answer:* factoring   **31.** 4 sec

32. $\left(9 + \sqrt{105}\right)$ sec (about 19.25 sec)   **33.** Earth

Reteaching with Practice

1. $8y(3y^2 + 4)$ **2.** $6n^3(n^5 - 3)$

3. $3(a^2 + 10)$ **4.** $2y(y + 3)(y - 3)$

5. $7t^3(t + 1)^2$ **6.** $x^2(x - 1)(x - 2)$

7. $(y^2 - 2)(y + 3)$ **8.** $(x^2 + 5)(x + 2)$

9. $(d^3 + 1)(d - 1)$ **10.** $-1, 5$ **11.** $0, 5, -5$

12. $-3, 0$

Real-Life Application

1.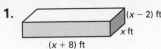

2. $96 = (x + 8)(x)(x - 2)$; $x = 4$

3. 12 ft by 4 ft by 2 ft   **4.** $142.22

5. 15 ft by 6 ft by 2 ft   **6.** No

Challenge: Skills and Applications

1. $(x - 6)(x - 1)(4x - 27)$

2. $(x^4 + 16)(x^2 + 4)(x + 2)(x - 2)$

3. $4(x + 2)(x - 7)$ **4.** $(8x + 9)(x + 3)(x - 3)$

5. $-4, -\frac{9}{2}$ **6.** $0, \frac{3}{2}, -\frac{3}{2}$ **7.** $0, -\frac{1}{4}, -\frac{11}{2}$

8. $3, -3$ **9. a.** -2 **b.** $2, 3, -5$

10. $\pi(r - 2)^2$ **11.** 2 meters   **12.** 12 meters

Review and Assessment

Review Games and Activities

1. C   **2.** A   **3.** N   **4.** O   **5.** P   **6.** E   **7.** A

8. S   CAN O PEAS

Test A

1. $5x^2 + 7x + 4$ **2.** $2x^2 + 3x + 3$

3. $9x^2 + 8x + 10$ **4.** $4x^2 + 3x + 1$

5. $-0.0002x^2 + 26x - 120,000$ **6.** $6x^2 - 10x$

7. $x^2 + 5x + 6$ **8.** $2x^2 + 7x + 3$

9. $x^3 + 2x^2 + 2x + 1$ **10.** $x^2 - 4$

11. $x^2 + 10x + 25$ **12.** $x^2 + 2x + 1$

13. $x = -4$ and $x = 2$ **14.** $x = -3$ **15.** B

16. C   **17.** A   **18.** $(x + 2)(x + 1)$

19. $(x - 3)(x + 2)$ **20.** $(x - 5)(x - 2)$

21. $(x - 3)(x + 4)$ **22.** length: $x + 3$; width: $x + 2$ **23.** $(2x + 3)(x + 1)$

24. $2(3x + 1)(x + 2)$ **25.** $(x + 3)(x - 3)$

26. $(x - 4)^2$ **27.** $(x + 3)^2$ **28.** $2x(x - 2)$

29. $(x^2 - 3)(x + 2)$ **30.** $(x^2 + 1)(x + 4)$

31. $0 = x^2 - 5x + 6$ **32.** $0 = x^2 - 4x + 3$

33. $x = -1$ and $x = 3$ **34.** $x = -3$ and $x = -2$ **35.** $x = -4$ and $x = 1$

36. $x = 1$ and $x = 5$ **37.** $x = -\frac{1}{3}$ and $x = -\frac{1}{2}$

38. $x = -3$ and $x = -\frac{3}{4}$

39. $x = -6$ and $x = 6$ **40.** $x = -6$

Test B

1. $x^2 + 7x - 2$ **2.** $3x^2 + 11x + 3$
3. $17x^2 + 4x - 4$ **4.** $11x^2 + 3x + 6$
5. $-0.0003x^2 + 42x - 110{,}000$
6. $24x^3 + 12x^2$ **7.** $x^2 - 9x + 20$
8. $4x^2 + 3x - 10$ **9.** $x^3 - 2x^2 - 2x - 3$
10. $x^2 - 25$ **11.** $x^2 - 16x + 64$
12. $2x^2 + 11x + 12$ **13.** $x = -5$ and $x = 1$
14. $x = -\frac{1}{2}$ **15.** B **16.** C **17.** A
18. $(x + 6)(x + 5)$ **19.** $(x - 8)(x + 5)$
20. $(x - 7)(x - 2)$ **21.** $(x + 7)(x - 5)$
22. length: $x + 4$; width: $x + 3$
23. $(5x + 1)(x + 3)$ **24.** $2(3x + 1)(x + 2)$
25. $(x + 4)(x - 4)$ **26.** $(x - 6)^2$ **27.** $(x + 5)^2$
28. $x(3x^2 + 9x + 2)$ **29.** $(x^2 + 2)(x - 3)$
30. $(x^2 + 5)(x - 1)$ **31.** $0 = x^2 - 8x + 15$
32. $0 = x^2 - 18x + 77$ **33.** $x = -2$ and $x = 4$
34. $x = -5$ and $x = -4$
35. $x = -6$ and $x = 2$
36. $x = 5$ and $x = 6$ **37.** $x = -\frac{1}{4}$ and $x = -1$
38. $x = -\frac{3}{2}$ and $x = -1$
39. $x = -7$ and $x = 7$ **40.** $x = 4$

Test C

1. $-x^2 + 5x - 20$ **2.** $10x^2 - 6x + 9$
3. $12x^2 - 8x + 1$ **4.** $17x^2 - 23x + 5$
5. $-0.0004x^2 + 40x - 130{,}000$
6. $-36x^3 - 12x^2 + 20x$ **7.** $x^2 - 4x - 32$
8. $27x^2 - 21x - 20$ **9.** $2x^3 + 2x^2 - 2x + 4$
10. $4x^2 - 9$ **11.** $25x^2 + 20x + 4$
12. $3x^2 - \frac{7}{2}x - \frac{3}{2}$ **13.** $x = -4$ and $x = 2$
14. $x = -3, x = 1,$ and $x = 3$ **15.** B **16.** C
17. A **18.** $(x + 8)(x + 7)$ **19.** $(x - 9)(x + 7)$
20. $(x - 6)(x - 2)$ **21.** $(x + 7)(x - 4)$
22. length: $x + 2$; width: $x - 5$
23. $(3x - 1)(x + 2)$ **24.** $2(5x - 3)(x + 1)$
25. $(2x + 5)(2x - 5)$ **26.** $(2x - 5)^2$
27. $(3x + 7)^2$ **28.** $3x(2x + 1)(x - 3)$

29. $(x + 2)(x - 2)(x + 3)$
30. $(x + 4)(x - 4)(x + 5)$
31. $0 = x^2 - 3x - 10$ **32.** $0 = x^2 + 8x + 15$
33. $x = -3$ and $x = 5$ **34.** $x = -6$
35. $x = -8$ and $x = 5$ **36.** $x = 3$ and $x = 6$
37. $x = -\frac{1}{3}$ and $x = 1$ **38.** $x = -\frac{4}{9}$ and $x = 1$
39. $x = -\frac{5}{4}$ and $x = \frac{5}{4}$ **40.** $x = -1$

SAT/ACT Chapter Test

1. A **2.** B **3.** A **4.** D **5.** C **6.** A **7.** B
8. A **9.** C

Alternative Assessment

1. a–c. Complete answers should address these points. **a.** • Equations *i* and *ii* each have 2 solutions and *iii* has no solution. **b.** • Explain that problem *i* can be solved by all three methods, *ii* and *iii* can only be solved by quadratic formula and graphing. Explain reasons why. • Explain that *i* can be easily factored since its discriminant is a perfect square; *ii* cannot be factored but quadratic formula could be used to solve.
• Explain that in *ii* graphing would be harder to obtain an exact answer. • Explain that part *iii* is most easily solved by graphing because one can quickly tell that there are no *x*-intercepts.
c. • Explain advantages and disadvantages including exact versus approximate answers for graphing, factorability and how easy or difficult a problem may be to factor.

2. a. Figure 1 perimeter: $4x - 16$ units; Figure 2 perimeter: $3x^2 + 5x - 8$ units **b.** 4 cm; -4 cm which is not possible. *Sample Answer:* One must make sure that the lengths of the sides of the figure are positive numbers. **c.** 12
d. Solutions for x are $\frac{4}{3}$ and -3, however, neither is true because the length of one or more sides of the triangle would be negative. **e.** Figure 1 area: $x^2 - 8x + 16$ square units; Figure 3 area: $2x^2 - 17x + 36$ square units **f.** 66 cm²
g. 6 (x cannot be 2, otherwise sides would be negative) **h.** 5 (x cannot be 4, otherwise sides of square would be 0)

3. *Sample Answer:* To use zero-product property, equation must equal zero.

Review and Assessment *continued*

Project: Is it Real?

1. Check student's computations. *Sample answer for a volume:* 0.0575 liters

2. Make sure students give a reasonable estimate for the height of the doll or action figure if it were a real person. Check students' computations for both the scale factor and the volume. *Sample answer:* for a figure height of $4\frac{5}{16}$ in., an estimated height of 62 in., and the sample answer in Exercise 1, $s \approx 14.4$ and $V_p \approx 171.1$ liters (using the rounded value of 14.4 for s)

3. Check student's computations. *Sample answer:* Based on answers for Exercise 2, male weight ≈ 402 lb, female weight ≈ 397 lb; no

Cumulative Review

1. $10 + x$ **2.** $\frac{1}{3}x$ **3.** $8 - x$ **4.** $\frac{5^3}{x}$ **5.** -26

6. $-2{,}187$ **7.** -448 **8.** -5

9. Yes. Domain is 1, 2, 3, 4. Range is 8, 9, 13, 20. **10.** Yes. Domain is 2, 3, 4. Range is 2.

11. Yes. Domain is 10, 12, 18, 22. Range is 2, 3, 5.

12. $y = -\frac{4}{3}x + 12$

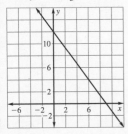

13. $y = 9$

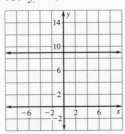

14. $y = -\frac{3}{5}x - \frac{19}{5}$

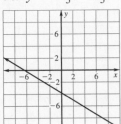

15. $y = -\frac{3}{2}x - \frac{1}{2}$

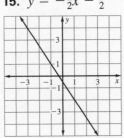

16. $y = -\frac{2}{3}x + 8$

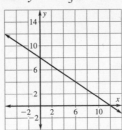

17. $y = -\frac{5}{3}x - 13$

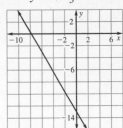

18.

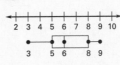

19.

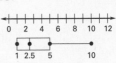

20.

21. $\left(\frac{44}{21}, \frac{73}{21}\right)$

22. $\left(\frac{7}{16}, \frac{43}{8}\right)$ **23.** $\left(-\frac{130}{11}, \frac{116}{11}\right)$ **24.** $\frac{x^2}{4}$

25. $\frac{1}{x^{45}y^{27}}$ **26.** $\frac{3y^8}{2x^2}$ **27.** r^{17} **28.** m^{18}

29. $\frac{2}{5x^{18}}$ **30.** $6 \pm \sqrt{35}$ **31.** $\frac{15 \pm \sqrt{205}}{2}$

32. no solution **33.** $-1, \frac{5}{2}$ **34.** no solution

35. one solution **36.** two solutions

37. $20x^4 + x + 1$ **38.** $-6t^3 - t^2 - t - 10$

39. $x^4 - \frac{4}{3}x^2 - 1$ **40.** $0.57m^4 - 20.9m$

41. $3r^2 - 2r - 8$ **42.** $6y^2 - 19y - 7$

43. $6t^2 + 3t - 18$ **44.** $28 - 26x + 6x^2$

45. $2.2q^2 - 8.3q - 2$ **46.** $s^2 - \frac{1}{4}s - \frac{1}{8}$

47. $x^2 + 12x + 36$ **48.** $4y^2 - 4y + 1$

49. $16s^2 + 16s + 4$ **50.** $x^2 - 4.6x + 5.29$

51. $\frac{1}{16}x^2 - \frac{1}{2}x + 1$ **52.** $25w^2 - 12w + 1.44$

53. $-1, \frac{2}{3}, \frac{5}{2}$ **54.** $\frac{1}{8}, \frac{1}{2}, \frac{3}{4}$ **55.** $0.4556, 0.7714$

56. $-6, \frac{8}{81}, 2$ **57.** $(2n + 1)^2$

58. $(4x + 7)(4x - 7)$ **59.** $(2x - 3)(x + 3)$

60. $(3x + 1)(2x + 2)$ **61.** $(y - 3)(3y + 1)$

62. $\left(\frac{1}{2}x + 3\right)\left(\frac{1}{2}x - 2\right)$ **63.** $6(4x^2 - 5)$

64. $7(a + 2)(a - 2)$ **65.** $9u(2u + 1)$

66. $x^2(-x^2 + 5)$ **67.** $(t + 9)(t - 9)(t - 1)$

68. $(c^3 + 1)(c + 2)$

Answers